Towards a Philosophy of Digital Media

Alberto Romele • Enrico Terrone
Editors

Towards a Philosophy of Digital Media

palgrave
macmillan

Editors
Alberto Romele
Université Catholique de Lille
Lille, France

Universidade do Porto
Porto, Portugal

Enrico Terrone
Università di Torino
Torino, Italy

ISBN 978-3-319-75758-2   ISBN 978-3-319-75759-9   (eBook)
https://doi.org/10.1007/978-3-319-75759-9

Library of Congress Control Number: 2018934728

Cover credit: Alexey Borodin

Printed on acid-free paper

This Palgrave Macmillan imprint is published by the registered company Springer International Publishing AG part of Springer Nature.
The registered company address is: Gewerbestrasse 11, 6330 Cham, Switzerland

Contents

Notes on Contributors

Bruno Bachimont is Professor at the Compiègne University of Technology-Sorbonne Universités (UTC) in France, where he teaches computer science, logic, and philosophy. He was Head of research of UTC from 2006 to 2017 and of the French National Audiovisual Institute (INA) from 1999 to 2001. He has published widely in the fields of artificial intelligence, knowledge-based systems, indexation, and document engineering. He is the author of "Le sens de la technique: le numérique et le calcul" (in French, Les Belles Lettres, 2010), and more recently of "Patrimoine et numérique: technique et politique de la mémoire" (in French, Ina-Éditions, 2017).

Jacopo Domenicucci is a PhD student at the University of Cambridge. He is a former student of the École Normale Supérieure of Paris. He has been studying trust since 2014 and digital trust since 2015. He has coedited a book on digital trust with Milad Doueihi, the volume being *La confiance à l'ère numérique* (in French, Éditions Berger-Levrault and Editions Rue d'Ulm, 2017).

Maurizio Ferraris is Professor of theoretical philosophy and Deputy Rector for humanities research at the University of Turin, where he is also the President of the research centre Laboratory for Ontology (LabOnt). He has written more than 50 books, translated into several languages, and is known for the theory of Documentality and for contemporary New Realism.

Fanny Georges is Assistant Professor in communications sciences at the Sorbonne Nouvelle University (Paris, France), in the Medias, Cultures and Digital Practices (MCPN/CIM) research team. She studies digital communication and more specifically digital identity, using a pragmatic semiotic approach. She was the Principal Investigator of the project "Digital eternities, post mortem identities and new memorial uses of the web from a gender perspective," funded by the French National Research Foundation (ANR) from 2014 to 2018.

Stefano Gualeni trained as an architect, is a philosopher and videogame designer best known for creating the videogames "Tony Tough and the Night of Roasted Moths" (1997), "Gua-Le-Ni; or, The Horrendous Parade" (2012), and "Something Something Soup Something" (2017). He works at the Institute of Digital Games of the University of Malta and is a Visiting Professor at the Laguna College of Art and Design (LCAD) in Laguna Beach, California. His work takes place at the intersections between continental philosophy and the design of virtual worlds.

Stacey O'Neil Irwin is Professor in the media and broadcasting program at Millersville University of Pennsylvania, USA, where she teaches creative and theoretical courses in digital media theory and production. Her research interests include digital media, media literacy, postphenomenology and philosophy of technology. She has authored work in *Phenomenology and Practice*, *Explorations in Media Ecology* and *Techné: Research in Philosophy & Technology*. She is the author of *Digital Media: Human-Technology Relations* (2016), and co-edited *Postphenomenology and Media: Essays on Human-Media-World Relations* (2017), both with Lexington Books.

Virginie Julliard is Assistant Professor in Information and Communication Sciences at the Compiègne University of Technology-Sorbonne Universités (France). Her work focuses on the media production of gender, digital writing apparatuses, and public debate. Her approach is socio-semiotic and semio-pragmatic. She wrote the book *De la presse à internet. La parité en questions* (in French, Hermès Lavoisier, 2012).

Jos de Mul is a professor in the Faculty of Philosophy at the Erasmus University Rotterdam, Netherlands, where he holds a chair in Philosophy of Man and Culture. He has also taught at the University of Michigan (Ann Arbor, 2007–2008), Fudan University (Shanghai, 2008), and Ritsumeikan University (Kyoto, 2016). In 2012, he became a visiting fellow at the Institute for Advanced Study in Princeton, USA. His research is on the interface of philosophical anthropology, philosophy of technology, aesthetics, and the history of nineteenth- and twentieth-century German philosophy. His books in English include *The Tragedy of Finitude: Dilthey's Hermeneutics of Life* (2004, 2010), *Cyberspace Odyssey: Towards a Virtual Ontology and Anthropology* (2010), and *Destiny Domesticated: The Rebirth of Tragedy out of the Spirit of Technology* (2014).

Janne Nielsen is Assistant Professor in media studies and Board Member of the Centre for Internet Studies at Aarhus University. She is part of DIGHUMLAB, where she is Head of LARM.fm (a community and research infrastructure for the study of audio and visual materials) and part of NetLab (a community and research infrastructure for the study of internet materials). She holds a PhD in media studies for her work on the historical use of cross media in the education of the Danish Broadcasting Corporation. Her research interests include media history, web historiography, web archiving, and cross media.

Alberto Romele is Researcher at the ETHICS Lab of the Lille Catholic University (France), and at the Institute of Philosophy of the University of Porto (Portugal). He has published articles in English in journals such as *Theory, Culture & Society*, *Surveillance & Society*, and *Techné: Research in Philosophy and Technology*. His researches focus on technological imaginaries, hermeneutics, and theories of the digital.

Marta Severo is Associate Professor in Communication at the University Paris Nanterre. She is a member and Deputy Director of the Dicen laboratory. Her research focuses on action research in social sciences through web-based data. Her activities concern two thematic areas: the collaborative construction of cultural heritage and memories on the web; and the digital representations of place and space. Since 2012, she has coordinated the research program "Media and territories" at the International College of Territorial Sciences (Paris).

Jacek Smolicki is a cross-media practitioner, researcher, and walker interested in the poetics and politics of memory practices. Working at the intersection of art, technology, archiving, and everyday life, his practical work comprises such forms as interactive installations, sousveillance art, soundwalks, site-specific memorial art, immersive soundscapes, and performances. In parallel to these activities, for the last several years Smolicki has been committed to a set of aesthetic practices in which he employs various recording technologies to construct an experimental para-archive of contemporary everyday life. Read more on www.smolicki.com

Enrico Terrone is a postdoctoral researcher at the Università di Torino and Aassociate Researcher at the Collège d'études mondiales, Paris. He was awarded a fellowship in Bonn (Käte Hamburger Kolleg) and another in Paris (FMSH—Gerda Henkel Stiftung). His work ranges over aesthetics, social ontology, and the philosophy of technology. His primary area of research is the philosophy of film.

Galit Wellner is a Senior Lecturer at the NB School of Design (Haifa, Israel). She is also an Adjunct Professor at Tel Aviv University. Galit studies digital technologies and their inter-relations with humans. She is an active member of the Postphenomenology Community. She has published several peer-reviewed articles and book chapters. Her book *A Postphenomenological Inquiry of Cellphones: Genealogies, Meanings and Becoming* was published in 2015 by Lexington Books. She translated into Hebrew Don Ihde's book *Postphenomenology and Technoscience* (Resling 2016).

List of Figures

List of Tables

Introduction

Alberto Romele and Enrico Terrone

Media can be characterized as artifacts whose primary function consists in widening our experiential and cognitive horizons. Without media, our experience and cognition would be limited to what we can directly perceive in our environment and then possibly recall from our memory. At most, this could be enriched by what we can imagine; that is, conjectures about the future or thoughts about possible state of affairs. However, without media, our experiential and cognitive horizon would remain rather narrow. It is only when media come into the picture that we start acquiring *mediated* information about parts of the world that are beyond the reach of our *direct* perception, of our memory and of our imagination.

Media are so intrinsic to our nature that it seems legitimate to wonder if it really makes sense to speak of an immediate relation of human beings

A. Romele (✉)
Université Catholique de Lille, Lille, France

Universidade do Porto, Porto, Portugal

E. Terrone
Università di Torino, Torino, Italy

© The Author(s) 2018
A. Romele, E. Terrone (eds.), *Towards a Philosophy of Digital Media*,
https://doi.org/10.1007/978-3-319-75759-9_1

1

with the world. Language is the medium that mostly characterizes us, which allows abstraction and communication. It is not surprising, then, that philosophers have devoted, especially during the twentieth century, much of their energies in order to understand what it is and how it works. Still, there are many other media. Pictures, sculptures and theatrical performances are, for instance, mediators of a specific kind. Philosophers have often circumscribed such media to the domain of the aesthetics and the philosophy of art. In fact, while the philosophy of language is clearly distinct from the philosophy of literature, one finds it hard to draw a similar distinction between the medium and its artistic use for what concerns pictures, sculptures or theatrical performances. As a consequence, most of philosophical reflections on such media have been conducted within the field of aesthetics (see for instance, Davies 2003; Lopes 2009; Gaut 2010). Yet, this approach reveals its limits once technology is considered.

The point is that technology can create new media by supporting other forms of media and enhancing their capacity to widen our horizon, regardless of their aesthetically relevant applications. For instance, printing supports language and depiction, thereby creating a new medium (the press) through which texts and pictures are widely spread by means of technical reproduction. Likewise, technologies such as the telegraph and the telephone empower language and linguistic communication.

On the one hand, there is a long tradition in philosophy of technology and in media studies that considers technologies as extensions of human bodies and minds. Ernst Kapp, who actually coined the term "philosophy of technology" in 1877, described tools and weapons as "organ projections." Notably, Marshall McLuhan presented media as "an extension of man." On the other hand, postphenomenologists (Ihde 1990; Verbeek 2011), as well as scholars like Friedrich Kittler (1999) and Bruno Latour (1994), insist on the fact that technologies are mediators that alter our relationship with the world in a fundamental and irreversible way. From this perspective, it is more than just the empowerment of human bodies and minds. Rather, it is the ability to produce brand-new relations with the world, thereby reshaping our own nature. The same tradition has also accounted for limit-cases in which technologies are not mediating anymore between humans and the world, but become autonomous entities. This is the case of Don Ihde's "alterity relations" or Latour's "delegation." This attitude has been taken to extremes by authors who, like Gilbert Simondon, have tried to follow the evolution of technological lineages independently of their interactions with or their relevance for human beings.

For sure, one can still understand all this from an aesthetic perspective. And yet, it is clear that this is just one aspect of what should be an encompassing philosophical perspective on media. Philosophers have been particularly attracted by the social and political consequences of the strong asymmetry characterizing mass media; namely, media whereby a few agents can supply information to a large majority of people. In their research at the frontier between aesthetics and political philosophy, Adorno and Horkheimer (2007) have criticized the culture industry of cinema precisely for allowing a small minority of subjects to supply imaginative variations to the masses, depriving viewers of the possibility of actively exercising their own imagination. Nonetheless, it seems fair to say that during the twentieth century philosophers have paid little attention to what these media actually are, and how they reconfigure our relationship to the world, over and above their political or aesthetic impact.

At the turn of the century, the rise of digital media significantly changed the situation. Existing media such as press, phonograpy, photography and cinema started digitalizing. That is to say that texts, sounds and pictures are no longer recorded as analog traces but rather as sequences of bits. Digital media became the most important interfaces between the world and us, "metamedia" capable of simulating the specificity of older media (Manovich 2013). Thanks to them, the production and reproduction of texts, sound recordings, pictures and videos have become easier and more widespread. Moreover, digital media have introduced new ways of widening our horizons. E-mails, the Web and social media are not just extensions and modifications of other media. By embedding preexisting media such as writing, depiction and sound recording, they have profoundly transformed them. And they have also altered our relationship to the world, to the others and to ourselves. Finally, digital media are henceforth considered more than just mediators: many of them are also seen and treated as valuable interlocutors.

Three main elements characterize digital media. These same elements also characterize the recent history of their interpretation and understanding by scholars from different backgrounds. We will label these three elements: (1) "interaction"; (2) "recording"; and (3) "autonomy."

1. First, digital media somehow subvert the traditional notion of mass media. In digital media, the masses do not passively receive information; rather, they contribute to its creation and diffusion. Individuals are no longer mere consumers, but also producers of contents—"prosumers." Furthermore, people are not isolated from their peers anymore; they can communicate with each other, they can interact. In this sense, digital media may appear as means of liberation of the masses from the few. Let us consider, for instance, the several publications on the gift economy online (Romele and Severo 2016) or those on the hacker ethics against the spirit of capitalism (Himanen 2001). Eventually, the virtual environments allow people to free themselves also from the physical and social constrictions in real life. This was at least the opinion of several digital pioneers who during the late 1980s and early 1990s imagined a utopian virtual world where sex, race, class, gender, and sexual orientation ceased to be relevant. Nowadays, things look significantly different. However, it remains true that digital media make room for much more action and interaction than traditional mass media.

2. Second, digital media involve a complete overlapping between communication, on the one hand, and recording, registration and keeping track, on the other. For sure, information production and communication remain important aspects of digital media (Floridi 2014). And yet, for a few years now, recording, registration and keeping track have taken the upper hand. For this reason, most of the chapters of *Towards a Philosophy of Digital Media* are devoted to this topic.

   In fact, the birth and development of the social Web at the beginning of the new millennium have greatly increased the amount of information concerning users that is recorded and accessible online. As a consequence, people and researchers have started to consider social media as technologies of mutual control and surveillance. Moreover, most of the Internet platforms have started a massive collection and analysis of prosumers' data (big data) in order to analyze and even anticipate their preferences. In 2013, the British newspaper the *Guardian* famously revealed that the US National Security Agency (NSA) obtained access to the systems of Google, Facebook, Apple and other US technology and Internet companies. In this general atmosphere of recentralization of the Web and new "vertical" surveillance, social media and, more generally, digital media have started to be treated like forms of post-Foucauldian Panopticons (Andrejevic and Gates 2014). However, for us, the question goes deeper, as far as it does not concern just ethics and politics. Digital recording, registration and keeping track have ontological, epistemological and anthropological implications.

From an ontological point of view, they have determined a significant overlapping between communication and registration. As a consequence, the question arises whether it is legitimate or not, and eventually to what extent and under which conditions, to understand the web as a series of traces, inscriptions or documents. Can we interpret the web as an archive? Would it be possible to define digital media as "recording(-based) media"?

From an epistemological point of view, digital traceability became a "total social fact." As a consequence, several scholars have seen in digital traces, their exploitations and treatments, not a danger, but rather an opportunity for filling the gap between natural and social sciences. On the one hand, it has been observed that actions and interactions online are in perfect continuity with those offline. Thus, we should rather talk of the "end of the virtual" (Rogers 2013), which means that digitally recorded actions and interactions are reliable representations of the social reality itself. On the other hand, it has been said that in contrast to the classic data produced by sociologists, digital traces are not produced in artificial situations. Rather, "they are generated as a by-product of already occurring interactions and processes across social life" (Marres 2017, 17). At present, one can notice "conflicts of methods" in this emerging field: digital sociological research, computational social science, digital methods, digital sociology, and cultural analytics are just a few of the labels that researchers use to name their approaches, and to differentiate them from those of other researchers.

From an anthropological point of view, digital recording, registration and keeping track have deeply changed the previous relation between remembering and forgetting. Certainly, one must not exaggerate by saying that everything digitally recorded and eventually posted online is there forever. For sure, many of you have experienced an irremediably broken hard-drive or losing an important digital document. Similarly, many of you have had to deal with linkrot, the process by which hyperlinks become permanently unavailable. Still, it is fair to say that while forgetting was the rule before the arrival of digital media, and remembering the results of an explicit decision, today it is the other way around—and this is why we are tragically lazy in backing up our digital data. This is so true that someone has proposed to implement an expiration date for the digital files that must be voluntarily decided by the one who produces or uploads them (Mayer-Schönberger 2011).

3. The third element characterizing digital media is that a crucial role is assigned to software, which is not only able to transmit and store information, but also to manipulate it. Digital media do not limit themselves to help our minds or to interconnect us with other human subjects. In digital media, recording often involves the autonomous manipulation, combination and recombination of what is recorded. So there arises the legitimate question of whether digital media, or at least an emerging part of them, such as those implementing machine learning algorithms, should not be considered as members of the social world, nonhuman actants. Here is where the philosophical reflection on digital media crosses the philosophy of artificial intelligence (see Haugeland 1985; Dreyfus 1991). What is relevant here is, however, a "lesser" AI, which, despite some claims for the "master algorithm" (Domingos 2015), is less interested in creating a universal machine than in accomplishing specific tasks like blocking e-mail spam or face recognition.

Of course, *Towards a Philosophy of Digital Media* is not the first philosophical book questioning digital media. Nor does this book claim to have the last word on this topic. The title should be understood simply as the intention of drawing an outline of *a* philosophy of digital media, valid at least for the time being, and capable of anticipating some of the inevitable forthcoming changes. Its intention is to interpret digital media from a specific perspective, that of digital recording, registration and keeping track, and to test this paradigm in different contexts. Furthermore, we are aware of the fact that philosophy cannot survive alone in this field, but must be in continuous dialog with other disciplines, such as communication studies, digital humanities, media studies and game studies. That is why this book includes chapters by philosophers and scholars from these various disciplines.

The chapters in the first part of the volume, which is dubbed "Digital Media as Recording Devices," highlight in various ways the recording-based nature of digital media. Bruno Bachimont shows how technological innovations, due to Internet protocols, have introduced a primacy of recording over communication. Maurizio Ferraris treats the recording-based nature of digital media as the ending point of a historical process whose previous stages were the era of capitalism as described by Marx, and the era of communication and spectacle, illustrated by Debord. Janne Nielsen ponders how the Web can be recorded. She argues that, in spite of being recording-based, the Web is in fact affected by a significant evanescence. Jacek Smolicki discusses the "capture culture" in relation to Bernard Stiegler's concepts of mnemotechniques and mnemotechnologies. He points towards an alternative way of thinking concerning our practices of archiving.

The chapters in the central part of the volume, "Consequences of Digital Recording," consider how digital recording bears upon specific issues. Jos De Mul shows that digital recording enables new ways of thinking by examining the case of Wikipedia, focusing on its collective character, and wondering if it can be seen as a "hive mind." Jacopo Domenicucci considers the impact of digital recording on trust regarding relationships between subjects. Fanny Georges and Virginie Julliard focus on a peculiar consequence of digital recording; namely, the duration of digital data after the death of a user. Marta Severo explores the new possibilities offered by digital participatory platforms for creating living inventories of oral cultures while avoiding fossilization of traditional written documents.

The chapters in the final part of the volume, which is dubbed "Digital Media Beyond Recording," individualize new perspectives that are relevant for understanding digital media. Stacey O'Neil Irwin explores the "taken-for-granted" aspects of digital media and interrogates their non-neutrality through a postphenomenological perspective. Galit Wellner uses postphenomenology to analyze "algorithmic writing," the idea that texts written by machine-driven algorithms appear as though an actual person wrote them. As a consequence, one has to consider the autonomy and the agency of these algorithms; that is, the fact that digital media seems to be dynamic, productive and creative. Stefano Gualeni distinguishes between two forms of "doing" with and through digital media: "doing as making," and "doing as acting." The former is, for instance, the "doing" of the designer of a video game while the latter is the "doing" of the player. Alberto Romele sketches the outlines of a "third paradigm," following those of communication and registration, to understand digital media. He speaks of electronic imagination or "*e*magination," which does not contradict digital recording, but is rather an emerging property of it.

References

Adorno, Theodor W., and Max Horkheimer. 2007. *Dialectic of Enlightenment*. Stanford: Stanford University Press.

Andrejevic, Mark, and Kelly Gates. 2014. Big Data Surveillance. Introduction. *Surveillance & Society* 12 (2): 185–196.

Davies, David. 2003. *Medium in Art. In the Oxford Handbook of Aesthetics*. Oxford: Oxford University Press.

Domingos, Pedro. 2015. *The Master Algorithm*. New York: Basic Books.

Dreyfus, Hubert L. 1991. *What Computers Still Can't Do*. Cambridge, MA: The MIT Press.

Floridi, Luciano. 2014. *The Fourth Revolution: How the Infosphere is Reshaping Human Reality*. Oxford: Oxford University Press.

Gaut, Berys. 2010. *A Philosophy of Cinematic Art*. Cambridge: Cambridge University Press.

Haugeland, John. 1985. *Artificial Intelligence. The Very Idea*. Cambridge, MA: The MIT Press.

Himanen, Pekka. 2001. *The Hacker Ethic and the Spirit of Information Age*. London: Martin Secker & Warburg.

Ihde, Don. 1990. *Technology and the Lifeworld. From Garden to Earth*. Bloomington, IN: Indiana University Press.

Kittler, Friedrich. 1999. *Gramophone, Film, Typewriter*. Stanford: Stanford University Press.

Latour, Bruno. 1994. On Technical Mediation. *Common Knowledge* 3 (2): 29–64.

Lopes, Dominic. 2009. *A Philosophy of Computer Art*. London: Routledge.

Manovich, Lev. 2013. *Software Takes Command*. New York: Bloomsbury.

Marres, Noortje. 2017. *Digital Sociology*. Cambridge: Polity Press.

Mayer-Schönberger, Viktor. 2011. *Delete. The Virtue of Forgetting in the Digital Age*. Princeton: Princeton University Press.

Rogers, Richard. 2013. *Digital Methods*. Cambridge, MA: The MIT Press.

Romele, Alberto, and Marta Severo. 2016. The Economy of the Digital Gift. From Socialism to Sociality Online. *Theory, Culture & Society* 33 (5): 43–63.

Verbeek, Peter-Paul. 2011. *Moralizing Technology: Understanding and Designing the Morality of Things*. Chicago: The University of Chicago Press.

Part I

Digital Media as Recording Devices

Between Formats and Data: When Communication Becomes Recording

Bruno Bachimont

1 Introduction

Communication and recording did not wait for the digital age to bump into each other. It all began when we started to want to communicate remotely, and to preserve the contents of a communication through time. Communication through time refers to the problematic of memory: the possibility of retaining contents from time evanescence is at stake. Communication through space refers to the problematic of journeys, transport, and movement. These two issues mobilize the same instrument for their solution: the document, that is to say, the registration of an event that is required to be remembered or communicated.

Recording is the tool and the technology that make communication possible even when the latter does no longer occurs in the co-presence of the interlocutors. For a long time, we lived according to this obvious rule: when the communication happens in co-presence, recording is not useful;

B. Bachimont (✉)
Sorbonne Université, Paris, France
e-mail: bruno.bachimont@sorbonne-universite.fr

A. Romele, E. Terrone (eds.), *Towards a Philosophy of Digital Media*,
https://doi.org/10.1007/978-3-319-75759-9_2

as soon as one wishes to cross space and/or time, recording becomes the necessary mediation.

This evidence has been valid for a long time. When radio and television were invented, we learned to communicate remotely, yet within the same temporal framework shared by the sender and the receiver. Recording was still useless; the goal was rather to carry the message from one person to another, from one place to another. We learned to do it through the propagation of a signal that did not need to be fixed on a medium in order to be repeatable and transmissible. It was necessary, instead, that the transmission and reception contexts were simultaneous so that they could be shared, on the one hand, and allow interaction on the other (only in principle, however, as shown by the first age of radio and television, when listeners and viewers could not interact with the sender). In other words, as long as the transmission was in real time, registration was useless, because we remained in a situation of telecommunication.

An inversion occurred when technologies based on the Internet Protocol (IP) introduced packet switching. Indeed, the paradigm was reversed: whereas, until then, we used to communicate without recording, and the issue of recording was eventually raised after the communication, the IP imposed recording in the form of packets first, in order to communicate these same packets in a second stage. Instead of a wave propagation routing a signal without having to record it, we now have a registration that we want to carry from one place to another.

At the beginning of the Internet age, this paradigm shift was not noticeable because the Internet was mostly dedicated to data communication, rather than to that of contents like voice, television, radio and cinema, for which reception must be assured in real time, in quasi-simultaneity with the emission. But we know what happened next: the network improvements allowed real-time transmission, and hence this type of transmission was implemented for telecommunications, in particular for radio and television. This has been the technological paradox of the 1990s: a protocol such as the IP is not adapted for telecommunications at all, but this same protocol's technical progress has assisted a general transition of telecommunications into the digital.

If we now register to communicate, this implies that communication is conditioned by the technical choices of the recording: in addition to the format of the signal encoding, it is also necessary to take into account the formats of recording and those of the manipulation involved in these technologies.

The recording techniques and technologies have two sorts of effects on communication. First, they impact the contents' formatting. Second, they contribute to the progressive reduction of documents to data. Indeed, the digital is above all a binary coding which ensures the manipulability of contents. But the binary code in itself does not refer to any particular semantics; it is purely arbitrary. Everything depends on the way it is decoded. The role of the formats consists precisely in disciplining the expressiveness of the binary code in order to give it a meaning. For example, it ensures that a particular binary file encodes a video and not a text or a sound. The format is therefore the key aspect enabling contents to be digital. Moreover, since we are dealing with registrations, that is to say with objects that can be manipulated, copied, fragmented, etc., one will try to bring back the contents to a type of basic records, facilitating as much as possible their management, transmission, processing, and transformation. This is done by considering contents as data; that is, as elementary records.

Contents are analyzed and broken down into data, and these data express themselves through formats that prescribe feasible operations and the possible structuration of information. Communication is therefore subordinated to what formats allow, and transformed by the data manipulation.

In this chapter, my aim is to consider this twofold movement (from bits to formats, and from contents to data) to account for some of the theoretical and technical consequences it implies. But before I expand on this movement, I would like to clarify some terminology and concepts, in order to define what I mean by content, document, communication, and recording.

2   Communication and Recording

Although communication and recording have been interacting strongly for a long time, these two notions refer to different concerns. Communication, at least in its ideal-typical form, evokes a situation where people share the same place and the same moment for exchanging signals that they interpret in order understand each other. Speech is the privileged medium, but not the only one. The sharing of a meaning can also happen through non-verbal registers characterizing a specific situation and interaction. Communication is therefore in principle an inter-comprehension in co-presence.

Recording is a palliative. It is required in order to mitigate the absence of what is not anymore, of what has changed, and the absence of what is not there. Recording has a double function. First, as far as it is produced by an event of which it is the trace, it represents a proof and a testimony of it. It has an indexical relation with this event. As such, it can inform about this event and it can constitute a memory of it, since it is permanent. Its own existence attests to that of the past event; moreover, its content can inform us about the nature of that event, especially when this trace is produced voluntarily, as with administrative records, the archives, which collect information about the event to be remembered. Second, being permanent, records allow multiple consultations, distributed in space and time: their existence does not coincide with the evanescence of the event; they are static and out of time, allowing multiple events of consultation. As long as it is material, a record can also be reproduced and copied for multiplying the possible consultations. Allowing the free repetition of reading and the reproduction of the recorded object, registrations exceed the singularity of the event and its spatiotemporal uniqueness.

2.1   Content, Inscription, Document

If a registration is the trace of an event (that is, a proof of its happening), then it also has a content. It conveys a message informing about the nature of the event. As such, it must be interpreted. It is therefore necessary to consider the nature of recording as a content and to consider its conditions of interpretability.

Human beings, beings of flesh and blood, communicate with each other. According to Aristotle, the human becomes human only insofar as it becomes political, and interacts and exchanges within the city. It is on this condition that the humanity expresses and accomplishes itself: the city as political space is the space in which the human animal becomes rational (that is to say, human), at least if one follows the classic definition *zoon logon echon* (Manent 2010).

Communication is based on the contents' exchange. A content can be defined as a *semiotic form of expression* associated with a *material medium of manifestation*; through its physical materiality, the medium makes a content perceptible, while the semiotic form makes it interpretable. As a meeting point between a materiality and a semiotic code, a content is addressed to our senses to speak to our spirit (Bachimont 2017).

Contents are therefore material and have the physical properties of their medium: intangible contents do not exist. But a content cannot be reduced to its material medium: what makes it a content and not just a thing, an object, is that it carries a signification that is addressed to an interpreter. Surprisingly, what renders an object a content is not so much the fact that it is produced to be a content, but that it is understood as such, as the bearer of a meaning or a message that is addressed to someone. Thus, even objects that have not been shaped by anyone can have a semiotic status, become a content, and be interpreted as we interpret a text, a book. Let us consider Galileo's famous affirmation according to which there is a book of nature written in the language of mathematics. A semiotic form sending *us* a message, *expressing* a meaning to be interpreted is recognized in a material object. Such a form is beyond the mere perception of the physical objects, and it makes out of it a content.

A content refers us to the ensemble of our cultural objects, such as books, newspapers, television programs, and so forth. But it also refers us to the voice: its material medium of manifestation is sound, the vibrating air, and the semiotic form is that of speech, the spoken language in which the heard sound becomes for the listener an interpreted and understood message.

A content is always material. But it is not necessarily permanent. For this reason, the physical medium, which lends its materiality to the content in order for it to be perceptible, must be static; that is to say, stable in space (and time). The content must be fixed on a static medium in order to be permanent. Such a content is called inscription.

Inscriptions, which became permanent thanks to their medium, considerably modify the conditions of communication. Unlike what happens to evanescent contents such as speech, it is no longer necessary for the speaker and the listener to be together in spatial contiguity and temporal simultaneity. In the basic configuration of communication, the listener must indeed be in the same place (and at the same time) as the speaker in order to perceive and interpret what is said. If the listener no longer shares this contiguity and simultaneity with the speaker, she has no access to the content. The inscription allows speakers and listeners to be separated in space and time: communication is asynchronous and dislocated. But since the speakers and listeners no longer share the same context, it is necessary to enrich the semiotic form with additional information so that the listener (or the reader) has sufficient indications to interpret the content she remotely receives while the speaker is absent. This is why the inscription must meet specific writing and reading genres that I am going to call here "editorial genres." These genres allow the contents' meaning to be shared between speaker and listener, sender and receiver, author and reader. For example, in medical communication, a medical record is a codified way of recording medical information about a person. Such genre calls for a particular type of reading that the medical staff are supposed to master. Similarly, scientific articles and dissertation theses are types of writing referring to particular types of reading.

Once immersed into an editorial genre that codifies both writing and reading, the inscription becomes a document. All documents are inscriptions, but the reverse is not true: inscriptions are not necessarily documents. For example, inscriptions that nobody consults, such as video surveillance records, are not documents. They will only become so during their eventual exploitation.

2.2   Recording as Communication

As suggested above, communication does not systematically imply registration. As long as one is in a situation of co-presence, be it spatial or temporal, the mediation of recording is useless. Even in the absence of spatial contiguity, temporal simultaneity makes communication possible that still does not need recording, but only transmission. Telecommunication technologies have been developed to allow remote (*télé-*, at a distance) communication: the content produced by a speaker or a sender in general is transmitted, propagated to its receivers. This content is by definition a temporal content; that is to say, a content that exists only as temporal duration. The one who sends or receives a temporal content must synchronize her stream of consciousness with the stream of the perceived object. It is therefore communication in temporal simultaneity, and the interlocutors share the same time in order to express what they have to say: a conversation, the listening of a song sung, a word, the vision of a spectacle. Communication is not an exchange of contents that circulate as material objects from hand to hand, but rather consists in sharing the same temporal progression.

Since all contents, as has been said, are material, from a physical point of view a temporal content is a wave, the propagation of a signal within time. As a consequence, it is a matter of transmitting this wave through a suitable channel to reinstitute its temporal progression to the intended receivers. This is what was invented with radio and televisual transmission. Concerning the latter, it is noteworthy that how to transmit and receive a video signal was invented long before the ability to record it, with the first Ampex video recorders becoming available in the 1960s, some 20 years after the invention of television.

This historical reminder is important because it makes it possible for us to understand that communication is not a problem of registration but of transmission. The first thing at stake is to transmit, enabling interlocutors to share the same temporal situation. Recording is often impossible (as in the absence of videos for many years), sometimes unnecessary. Moreover, it is only after having transmitted that one raises the question of recording the content or not.

The computerization of communication networks, especially through the IPs and communication via packages, has upset this balance. The Internet was designed to transfer data, that is to say, records. It was not temporal simultaneity that was at stake, but the transfer of information, when the spatial contiguity between the transmitter and receiver was no longer provided. The registered object is then fragmented into packets which are routed along different channels, the communicated object being reconstituted only after the retrieval of these packets. The time of transfer is not a concern in this context; in particular, it is not relevant that the receiver receives the contents at the same time as their emission, and with the same rhythm and rate, as in a telephonic conversation.

These principles, which are totally distinct from those of communication, show that, a priori, the Internet is definitely not made to ensure the telecommunication of contents. But we know how the improved connection speed and bandwidth enabled packets to be reconstructed on arrival as if we were dealing with transmitted or propagated temporal contents. The incredible happened then, because recording allowed simulating communication.

The practical and theoretical consequences are immense. From a practical point of view, if before it was first necessary to communicate and then to raise the question of registration, now every communication presupposes a prior registration. As a result, we are faced with questions that did not make much sense before. For instance, should phone conversations be preserved? No effort is necessary to record them; but a decision must be taken about their preservation or not. Communication, in its constitutive temporal evanescence, is now fixed in digital recording and it has acquired the permanence that it previously lacked.

But we also understand that this shift from the time of communication to the space of recording, in which the temporal progression gives rise to a spatial organization of information packages, cannot be done without a particular mediation: that of coding. Indeed, a temporal stream cannot simply be stored to be repeated at a later date: only what is static and spatial can be stored, while the stream is dynamic and temporal. Coding makes it possible to translate the temporal progression into spatial organization, and to restore this progression from this organization at will. The code becomes the main operator for allowing the shift from communication to recording, on the one hand, and the simulation of communication from recording, on the other hand.

Being an object recorded on a permanent medium, the code is available for decoding. But since it is available and accessible, it can also be transformed, reconfiguring in this way the original content and modifying the final rendition. Some transformations of the code are inherent to the technical process (compilation, transcoding), which is aimed at maintaining the integrity of contents. Other transformations may deliberately be designed for transforming the content itself, and not just its code. This is the issue for editing. Thus, the first notable application of the digital in the audiovisual production was the virtual editing, allowing the manipulation of the code of the images in order to elaborate the reconfiguration of fragments (the rushes) and their assemblage.

The code itself is not a univocal entity. Rather, it is a binary sequence read according to specific conventions, namely the formats that condition the possibilities of decoding and manipulation. In other words, once communication becomes the decoding of a registration, it is submitted to the power of the formats.

3   From Bits to Formats

3.1   The Digital as a Calculation that Does Not Make Sense

If the technology is a matter of devices, the digital is a matter of manipulations (Bachimont 2010). The device arranges temporal progressions into spatial configurations; manipulation consists of combinations of primitive and elementary entities, entities that we propose to call *calculi*, following the etymology of the term "calculation." The digital is a calculating device, a device for manipulating *calculi*. These latter are abstractions of entities manipulated by devices of all kinds, and are represented and controlled by this same abstraction.

The digital is both a particular case of technological devices and what constitutes the very principle of any device. Indeed, since a device is the programming of a repetition and the spatial organization the programming of the temporal succession, then any device is a program. Digital technology is the culmination of technology in general, since it proposes an abstraction that applies to all technical devices and is itself the object of a technology of its own. It is for this reason that the digital is universal and is henceforth encountered in all technological fields, regardless of their nature. Every technology is a digital manipulation materialized and concretized, and contemporary digital technologies are doing no more than abstracting their underlying digital principle.

If one wants to consider the digital on its own principles/merits, two notions emerge: discretization and manipulation. Discretization means the fact that the technical object is related to an assembly of formal units; that is to say, entities defined solely by their formal properties so that they are mechanically distinguishable and without possible ambiguity. The meaning of these entities is reduced to their distinguishability: if one considers the binary alphabet, the one used most, the only thing that defines 1 is the fact of it being different from 0, and vice versa. Moreover, these assemblies of formal units are manipulated, transformed into other assemblies through rules which are themselves unambiguously and mechanically applicable.

As a result, it is clear that the digital is the rule of the nonsense: both formal units *(calculi)* and manipulation rules are meaningless and only consider the units according to their distinguishability: 0 as different from 1. In other words, the digital is based on the notion of type or category in order to determine the operations to be carried out on the assemblies: if we know that it is "1," then here is the transformation to perform.

But that is not all: since the formal entities are defined by their distinguishability, it is not necessary to specify how they are physically realized. The physical reality of the entities does not matter as long as they meet the requirements of distinguishability. The digital is indifferent to the material medium that realizes the formal assembly. Although digital manipulation cannot be immaterial, it does not depend on the material nature of the formal entities. Digital is then "independent from the medium," which is incidentally the definition in the digital context of what the term "virtual" means. From here stems the notion of virtualization, which means that a device is made independent from its material medium via the digital. Once digitized, it can be concretized by any medium that meets these requirements.

Sign Asceticism and *independence from the medium* are the two properties which derive from discretization and manipulation, these latter properties constitutively characterizing the digital. Asceticism, which etymologically means "exercise," refers here to two complementary things. First, to the fact that one should ignore the meaning to reduce any sign to a *calculus* allowing for a blind manipulation. Second, one should also learn to consider digital entities not as signs (signifiers that mean something) but as signifiers without signified, mere entities waiting for manipulation. In the context of the digital, all the art of an engineer consists of an exercise of abstention from meaning in order to propose a digital treatment. A difficult exercise for us human beings, who are above all semiotic animals that approach what surrounds us by its significance, as a message that we must interpret since the world is not merely reduced to what is shown here and now. Computer science is a spiritual asceticism of meaning.

3.2   The Digital: Anonymous and Wanderer

These two principles have fundamental consequences for the nature of digital or digitized contents. Indeed, they become *anonymous* and *wanderer*.

The digital, through this formal abstraction into meaningless symbols, becomes independent from the physical matter, which specifically realizes these symbols: *different* physical substrates can realize the *same* digital object. Not only are these substrates different hardware implementations (different *tokens*; for example, one video copied several times on a hard drive), but they can also be of different physical *types* in the sense that, according to the physical sciences, they can refer to different principles (optical or magnetic, to mention only the most important physical types that concretize the digital). Consequently, the same digital object becomes indifferent to the material that realizes it, and it can be realized in several different places, according to different physical principles while being the same. It is thus a wanderer: its free circulation from one physical realization to another does not alter its digital nature.

Since manipulation is meaningless, it is done independently from what the digital symbols are supposed to represent for us. What the machine does, has nothing to do with what we think it does. We think, for example, that it plays chess while it only handles 0 and 1. Digital content is then *anonymous*: it does not say anything in particular and does not refer to any particular interpretation; the same binary substrate can be read as a sound or as a video according to the reading format adopted.

These properties of wandering and anonymity, which make the digital nameless and stateless, proceed from what we have called the "double break" of the digital (Bachimont 2007); namely, a material break from the physical medium (the digital as a wanderer) and a semantic break from interpretation (the digital is anonymous).

3.3   Formats

If we refer just to the *calculi*, the digital is a totality closed in itself: anonymous and wanderer. We must therefore define the means allowing the digital to be something else, to refer to something else. This is the very principle of coding that we have already experienced, but which takes here the name of "format": if a digital code is the code of a particular content, the format is the method of performing the coding, on the one hand, and of organizing the information, on the other hand. The format generalizes the notion of code, and allows the digital to leave its splendid isolation.

A format can be defined as a writing and reading convention of formal assemblies with a view to a given functionality. These functionalities are of different kinds:

1. Coding format: encoding and decoding of contents, utterances, images, sounds, videos, texts, etc. This format connects the *calculus*, which is blind and anonymous, to something meaningful.
2. Programming language: programming of an execution and of an interaction; transformation of a content and interaction with a user. The programming language is what allows the programmer to address and control the manipulation of *calculi*, otherwise meaningless. It is the format for accessing the manipulation, what allows projecting a meaning (that of the programmer) on what is meaningless (manipulation).
3. Physical Formats: realization of formal entities (0 and 1) in a material substrate. This format is of highest importance since it makes the link between the *calculi*, which are abstract entities that individuate themselves only through manipulation, and the physical objects that turn abstracted calculi into material realities.

The format is thus what allows the digital world to emerge from its splendid isolation and to relate to the external world that it represents (what is encoded by the digital), to the physical world that realizes it, and finally the socio-technical world that uses it (the programmers).

The format has the peculiarity of allowing some use of the digital, but thereby prohibiting others. By making possible, the format prohibits. Faced with the binary abstraction, which can represent everything, because it represents nothing (see the *sign asceticism*), the format constrains the binary assembly in its constitution and exploitation in order to allow a given use, but makes other uses impossible. If a format makes it possible to find an image in a binary assembly realized via this same format, one will not be able to find an intelligible sound there. It will always be possible to decode the binary assembly as a sound, but one will get no more than noise, since no format can guarantee that we will find an intelligible sound from this assembly.

4   From Documents to Data

We have just seen that calculation, blind and formal manipulation, is dedicated to a specific use thanks to a format. This use is the one we found in the digital applications, in which, for example, the format allows coding or decoding of an image, a text, or a sound. The format allows a manipulation that corresponds to an operation on contents and that therefore has, from this point of view, a meaning. It can be noted that there is a similar movement concerning documentary contents. Indeed, data format documents in order to make them calculable. Formats disclose calculations and make them available for use by giving them a meaning, while data reduce documents to manipulable data.

4.1   Signs Interpreted, Manipulated Technical Units

This passage from document to data sets tension between two complementary dimensions of the document as a technical subject, on the one hand, and as a semiotic object, on the other. As a technical object, the document consists of manipulable entities, which help to build and transform it. These entities are based on what may be called technical units of manipulation, which are the smallest constitutive entities of the content, and are accessible to the technical system. For example, the digital text, defined as a stream of characters, adopts the character as a technical unit of manipulation (TUM). Digital photography is defined by the pixel, handled directly by photo processing applications, and graphic contents are defined by dots and so forth. These units are defined a priori and define the space of possibilities, of what is possible to do by manipulating them. Anything that cannot be defined as a combination of these units is impossible, any combination is however possible in principle. The choice of a particular TUM is a compromise between the complexity of the combinations to build and the benefits from the uses one wants to make possible. Thus, a word processor allows us to do everything in a document concerning characters; it is useless however to go to the level of pixels or dots, which therefore are not made accessible and manipulable by word processors. For doing this, we must use other tools, such as those of desktop publishing.

Defined a priori, constituting the basis of an associated technical system allowing their manipulation, TUMs format contents: they codify and standardize it in order to make it a space for possible manipulations.

This characterization of the document as a technical object differs from that which can be developed by considering its semiotic properties. As a cultural object, the document is interpreted: the issue is to determine what is significant in the document and what it means. In other words, the reader should find in the document what is meaningful according to her and what makes sense. However, as taught by hermeneutics (Grondin 1993) and contemporary semantic theories (based on the Saussurrian tradition in particular), signs emerged through interpretation and do not constitute an a priori condition. We must interpret the document to see what is significant in it, and so which are the signifiers; the sign is not defined before interpretation but it is rather its result. One can recognize here a holistic conception of meaning and language, where the whole precedes its parts and is their condition of existence. Not only the whole cannot be reduced to the sum of its parts, but the parts cannot be defined independently from the whole of which they are the components. The idea is now well established according to which the interactions within a system allow the parts to constitute and individuate themselves (Varela 1989; Simondon 2005). The interpretation thus goes from being global to local.

This is why the interpretation is always ambiguous. For a given interpretation, the signs are defined and so are their meanings; however, there are always several possible interpretations depending on the global context selected to individuate the parts of the discourse, the signs that constitute it. These signs are then semiotic interpretation units (SIUs); that is to say, saliencies carrying out the meaning that is freed through interpretation.

It is therefore understandable that the document is caught in a tension between signs defined a posteriori and technical units defined a priori. Indeed, the TUMs and SIUs follow two different approaches: the former are defined a priori and used to build the document editing tools and to manipulate contents; the latter emerge after an interpretive path and depend on the overall context in which documents and readers are immersed. There is no certainty that the TUMs and SIUs are identical or consistent with each other: what we can do in a document does not coincide by nature with what one understands.

For example, in a text document, we can manipulate characters. We can identify anything that is defined in terms of character. In particular, we can identify any string of characters separated by two white spaces, which is a reasonable approximation of what can be called a word from an interpretative and linguistic point of view. But any string of characters separated by white spaces does not signify a word, especially when you have poorly integrated morphemes or lexicalized expressions. Also, it is often difficult to individuate all the forms of the same word within a text.

This difference between technical manipulation and semiotic interpretation involves an ongoing work of translation between the two, where manipulation must formalize, model, and simulate interpretation. This is done well when the context is fixed, because, as we said, in this case there is no ambiguity. Since we already know what signs must be identified, modeling can simulate interpretation through manipulation. However, when the context is open, modeling is not possible, and the models that are proposed are therefore faulty. We all live this experience with the tools for spell-checking which, despite their formal sophistication, are often wrong, not because of their imperfection, but because of their ignorance of the writing context in which the document is produced.

4.2   Data as Coincidence of Manipulation and Interpretation

However, if the context is fixed and normalized, it is possible to reconcile technical signs and manipulable units to force the identity of those entities. The benefits would be considerable: manipulation would "make sense," and all that we understand could be translated into manipulation. This remarkable situation exists in the classical field of databases, and in the more recent contexts of the semantic web and linked data.

From this perspective, the expression is not just formatted, but also interpreted. The manipulation of the technical units takes a semiotic value leading to an interpretation that gives a meaning. However, manipulation and interpretation, taken together, constitute an *inference*. Inference is indeed a manipulation preserving meaning and truth: it is not an arbitrary manipulation whose meaning must be found a posteriori. It enables anticipation and prediction of which meaning the manipulation and its output will have for a reader or a user.

Hence, it is not only the material expression of the document that is formatted, but also its meaning. We can then define *information* as the formatting of contents, allowing not only the manipulation, but also the interpretation, and hence the inference. The *data* are then the smallest significant unit of information. Data are elementary registrations of a database, or the resource description framework (RDF) triple in the context of the Web of data.

The formatted document is no longer made of signs, which are more or less rebels against the interpretation that is in charge of discovering, identifying or even constituting them; it is rather a set of information resulting from data combination.

The formalization of document in information, or in other words, this formatting, is a gain and a loss. The gain is reflected in the fact that it is now possible to anticipate the meaning through the inferences: technical combination of data remains intelligible and interpretable. The loss consists in the fact that for meaning anticipation to succeed it must discipline the context to fix the interpretive horizons.

This simple observation leads to the conclusion: the formatting of meaning is only a useful instrument in particular contexts where the use is sufficiently stabilized to allow modeling interpretation by means of inference. Intelligibility can be modeled just locally.

5   Conclusion: Conditions for Intelligibility and Freedom to Interpret

The recent technological innovations, in particular, since IPs have been developed, have introduced a primarity of recording over communication, and have therefore given importance to formats. Indeed, recording consists of inscribing contents on a permanent material medium, and the format is what establishes correspondence between the content of the expressive unit and the technical unit of inscription on the medium.

The turmoil that we have experienced since then is mostly because recording is performed on a computational substrate, where the coding discloses the universal possibilities of calculation. The format has since then taken a dual function: it is what allows computational coding to be opened on applications and to make some uses possible; it also reduces documents to a mere sum of information defined through data. Format turns bits into meaningful entities, from bits to format, but also reduces interpretation to standardized and fixed situations, from documents to data.

The question is to what extent these tools of manipulation and of meaning admit an exteriority where interpretative, cognitive, and social invention is created. In other words, are formatted documents simply powerful and structuring tools, or are they the horizon from which our thinking must be built? Is there still room for interpretive freedom?

Two answers can be given. The first says that human beings are by nature semiotic animals that individuate meaning depending on the dynamics of contexts and situations they live in. Therefore, interpretive freedom is inherent to human behavior. But, and this is the second possible answer, our technical systems are increasingly globalizing and inter-personalizing communication, and hence are normalizing it according to planned situations that are provided by the technological communication systems themselves. Therefore, the invention of everyday life and ordinary situations is being progressively reduced to being no more than one of the possible combinations that have already been constructed on the basis of the formatted data.

This question is not just philosophical, but also political. Let us think about the several current discussions on the governance of algorithms and databases (Rouvroy and Berns 2013; Cardon 2015). The debate is open, and it must be kept open.

References

Bachimont, Bruno. 2007. *Ingénierie des connaissances et des contenus. Le numérique entre ontologies et documents*. Paris: Hermès.

———. 2010. *Le sens de la technique. Le numérique et le calcul*. Paris: Les Belles Lettres.

———. 2017. *Patrimoine et numérique. Technique et politique de la mémoire*. Bry-sur-Marne: INA Editions.

Cardon, Dominique. 2015. *A quoi rêvent les algorithmes?* Paris: Le Seuil.

Grondin, Jean. 1993. *L'universalité de l'herméneutique*. Paris: Presses Universitaires de France.

Manent, Pierre. 2010. *Les métamorphoses de la cité. Essai sur la dynamique de l'Occident*. Paris: Flammarion.

Rouvroy, Antoinette, and Thomas Berns. 2013. Gouvernementalité algorithmique et perspectives d'émancipation. Le disparate comme condition d'individuation par la relation? *Réseaux* 177 (1): 163–196.

Simondon, Gilbert. 2005. *L'individuation à la lumière des notions de forme et d'information*. Grenoble: Jérôme Million.

Varela, Francisco J. 1989. *Autonomie et connaissance*. Paris: Seuil.

From Capital to Documediality

Maurizio Ferraris

1 Introduction

Analyzing the social reality of his time, Marx liked to make the example of the production and sale of fabric; unfortunately, we still use his categories now that the fabric is that of the Internet. In other words, we still have not understood what is, indeed, a veritable revolution (the third, after the industrial one and the media one) which I call the "documedia revolution." Documediality indicates the allegiance between the constitutive power of documents (documentality) (Ferraris 2012) and the mobilizing power of the media. I propose to outline these three great transformations as follows (Table 1).

Capitalism in the strict sense corresponds to the economic era of production, and to the political era of liberalism. Here, recording (capital, as we will see) surrounded the process, being at its beginning, as a premise, and at the end, as the profit obtained through the production of goods.

M. Ferraris (✉)
University of Turin, Turin, Italy
e-mail: maurizio.ferraris@unito.it

© The Author(s) 2018
A. Romele, E. Terrone (eds.), *Towards a Philosophy of Digital Media*,
https://doi.org/10.1007/978-3-319-75759-9_3

Table 1 From capital to documediality

Liberalism	Capital (production)
Populism	Mediality (communication)
Post-truthism	Documediality (recording)

Populism, where communication has the upper hand over production, corresponds instead to the phase of mediality, where memory was short and ephemeral like the words on the radio before streaming became available: in short, it was the system of the 15 minutes of fame. Finally, documediality corresponds to a third phase, characterized by recording: that is, by the use of a huge apparatus, namely the Web, which has the essential feature of keeping track of any interaction. Each category subsumes the previous ones at a more general level: communication sheds light on the mysteries of commodities (the fact that they constitute the solidification of social relationships and are essentially social objects); recording, in turn, manifests in its structure and functioning the formula Object = Recorded Act, which is the constitutive law of social acts. The social object (commodity, news, fake news, symbolic poem, title, and so forth) is the result of a social act (involving at least two people: a worker and an employer, a buyer and a seller, an author and a reader [. . .]) characterized by the fact of being recorded on some medium, including the mind of the two social actors. Let's briefly examine the three categories, before analyzing them in depth.

Production. Marx, the heir of a Faustian (and Fichtean) age that has little in common with ours, was fascinated by a fundamental force: work as a process that takes place between man and nature. The centrality of production corresponds to the *Homo faber* and the *Homo economicus*, determining a selective representation of the social world in which there is no space for phenomena (now highlighted by documediality) like unproductive waste. The latter is not an exotic or futuristic idea, described by eccentric sociologists or anthropologists, but the current social praxis: is there any greater waste than the careless words daily thrown around on the Web? In this sense, the law of commodity exchange would only be a variation of the constitutive law of social objects in which a society made up of at least two people produces objects which may (but do not necessarily have to) be commodities and may (but do not necessarily have to) produce profit for the one and sustenance for the other. The only irreplaceable features of this law are the fact that the act must take place between at least two people (or between a person and a delegated machine, such as when purchasing a ticket online) and that it must be recorded, for the object to acquire ontological consistency.

Communication. If the predominant activity of manufacturing was the production of artifacts, the predominant activity of mediality was communication; production survived, but in a subordinate form. The new phase did not eliminate the previous ones, but made them less important: a TV channel can produce shows, but is still a TV channel even if it merely broadcasts content produced elsewhere. In the communication phase, work was no longer a source of coordination and formation of behaviors and classes. This change occurred in the West in roughly the 1930s (in the sense of a prevalent character: if the First World War was a war of materials, the Second World War, in which more material was used, was mainly a war of ideology and communication). This turn was understood, albeit after some delay, in the second half of the twentieth century, and the new situation was often interpreted with categories borrowed from the previous phase, conceiving it as a superstructure with respect to a production mechanism. This gave rise to two diverging interpretations. The first, classically Marxist, saw the structure as reality and the superstructure as appearance; the second, postmodern, considered the appearance as the authentic (social and, in some cases, natural) reality. In both interpretations, however, the fundamental trait of communication, namely unidirectionality, was understood correctly: the message, generally centralized, went from the holder of the means of production or communication to the worker (who alienated his workforce) or to the consumer (who alienated his free time).

Recording. This situation changed with the documedia revolution: reception and transmission are now identified with social media, and do not generate artifacts in relation to nature, nor communicate messages that leave no trace, but produce documents; that is, social objects, bringing to light the structures that were implicit in the previous phases. In the third book of *Capital*, moving on from the analysis of work, Marx had already focused on the examination of credit and accounting mechanisms, proposing a theory of document capital: commercial capital and trade precede the capitalist mode of production, and money is a sui generis commodity (in fact, through the lenses of the documedia revolution, money and commodities are both documents). However, Marx saw money as a supplement to nature: it is based on international trade, just as the exchange value overlaps with an original and natural use value. In the documedia perspective, however, it is the supplement (document, money, recording) that constitutes the foundation of society, with a process coextensive with the birth of writing, religion, and society. If, in agreement with Marx, the use value of borrowed money is its ability to act as capital, then capital is revealed as recording, and documentality appears to be previous both to commercial exchange and to industrial production.

2   Documents, Not Commodities

This documentality becomes easier to read over time. The analysis of capital offered less than20 years ago by economist Hernando De Soto (2000) suggests that the necessary condition for its constitution is the presence of documents, which set values and make it possible to transfer assets that are otherwise related to their here-and-now (in plain words, without documents, it is impossible to prove the ownership of a house, and therefore to sell it or mortgage it by turning it into capital). So, if, according to the Marxist view, the public debt is what gives money the ability to "procreate" (or to become "Faustian," as Spengler would say), by laying the foundations for the development of capital, this is because both debt and money (and capital) are subordinate variables of documentality. Piketty's work on the prevalence of capital performance over economic growth (Piketty 2014), along with his equivalence between capital and wealth, confirms the priority of documentality over work. And it could be added that social interaction is not reduced to accounting books and asset registries, but includes the Code Napoléon, the Bible, the Koran, and the Library of Babel called the Web (Weber had already seen Protestantism as the origin of capitalism but had failed to consider that the origin of that religion was precisely the privatization of consciousness generated by the spread of the printed Bible). The mysterious nature of the commodity as something sensibly suprasensible is thus revealed in the succession that from the goods and its phantasmagoria leads to spectacle,

Table 2 From commodity to social object

Manufacture (production)	Commodity
Mediality (communication)	Spectacle
Documediality (recording)	Social object

and hence (with a generalization that manifests its conceptual essence) to social objects (Table 2).

Commodity. In the great narrative of factories and exploitation unfolding in the first book of *Capital,* industrial capital is presented by Marx in the terms in which Freud addresses adult sexuality: namely as the foundation against which any other sexuality is either anticipation or degeneration. So much so that (although Marx acknowledges that commercial capital and interest-bearing capital come before industrial capital) the standard should be sought in the latter and not in the former. As we have seen, only in the third book does Marx acknowledge that there is a function superordinate to land ownership and to commercial or usurer's capital: the document and the recording of which the document is a manifestation. There is no commodity before recording, because the commodity results from an exchange, and the exchange presupposes debts and credits, and therefore recordings (at least in the minds of the actors of the transaction).

Spectacle. In manufacture people act, in front of the television they stare and (as we will see) in front of the computer screen they act again, in a way that completes and perfects both the manufacturing phase (activity) and the media phase (relationship with the world mediated by technical interfaces). In this sense, interpreting spectacle as alienation, as many readings of the media society have done, seems to be simply (and unreasonably) nostalgic for the world of production. Seeing television as a form of brutalization means forgetting the descriptions of the illiteracy and degradation of the English working class that can be found in *Capital.* Conversely, it is worth remembering that the spectacle as a social object reveals the mysterious nature of the commodity, defining itself (writes Debord, by curiously focusing on the image, almost as if music or literature were not a spectacle) as a "social relationship between people mediated by images" (Debord 1994, paragraph 4).

Social object. So each category subsumes the previous to a more general level, and we have seen how this works with commodities. From the documedia viewpoint, the latter (and, even more obviously, spectacles) are indeed (to borrow Marx's famous words) "sensibly suprasensible things"; that is, social things. The relationship between things is but the social relationship between people, and things manifest themselves as documents. Now, the metamorphosis of the commodity described by Marx is the passage from a natural object (use value) to a social object (exchange value), which he interprets essentially as "X counts as Y in C"; a formula made familiar by contemporary social ontologies (Searle 1995), where a physical object X (for example, a piece of paper) counts as a social object Y (for example a banknote) in a given context. Instead, it should be interpreted more accurately as "Object = Recorded Act": a formula that explains why the growth of recording, caused by documediality, has generated a social revolution like the one we are witnessing.

3   Mobilization, Not Labor

Consider the ruthless mobilization of humanity described with much pathos in the nineteenth century; or the hours of idleness and passivity of the media age, as well as the numbing of consciences (at least, so it has been called, with an inappropriate and moralistic analysis) in front of the television screen described (with equal albeit less motivated pathos) ever since the 1950s. All these things have been replaced by a singular phenomenon in the age of documediality: a mobilization that is even more extreme than that dreamed of by nineteenth-century capitalists. Night shifts and child labor are now the norm, not an exception, and there is no legal or humanitarian procedure trying to change this. And yet this mobilization does not entail a financial compensation for the voluntary mobilized, who pay *themselves* for the means of production in a process where self-valorization coincides with self-exploitation (if such a thing is even possible) (Table 3).

Table 3 From labor to mobilization

Manufacture (production)	Labor
Mediality (communication)	Consumerism
Documediality (recording)	Mobilization

Labor. The free worker for Marx was the indispensable tool for transforming money into capital. This however contradicts what he claimed about the document-like nature of the latter (by which, recording is enough to generate money, the loan and eventually capital) and did not allow him to conceive of a non-working society. But that is exactly what we are witnessing today, for better or worse. Labor, of course, is still there, just like exploitation, but it is now a secondary phenomenon compared to a situation that would have been inconceivable in Marx's age: the difference between work time and life time, between alienation and reproduction of workforce, is now unrecognizable. On the one hand, there seems to be a total mobilization by which the worker's life is wholly alienated, with a colonization that appears to have realized (albeit softly) the dream of a 24/7 work time (it is virtually possible to receive a work email at any moment of the day or night). On the other hand, though, in a visibly contradictory way, there seems to be no trace of alienation, because the variety of tasks and the lack of a rigid schedule turn the worker (a name we still use for lack of better alternatives) into the full (albeit ironic) realization of labor in the communist society.

The same person who may be called to answer to an e-mail in the middle of the night may write a Wikipedia page or (which is more likely) update her status on a social network during the day. This can only be compared to the finally liberated world described in *The German Ideology*: one where (mutatis mutandis) today you update your status and tomorrow answer to an e-mail, travel low cost in the morning and write a critical essay in the afternoon, while in the evening you post the picture of the hamburger you are eating. And yet one shouldn't forget that, after referring to this multitasking life as the perfection of man, Marx disapproved of the fact that the steam machine was invented by a watchmaker, the car frame by a barber, and the steamboat by a goldsmith! Had he known that Bell wanted to create a radio and invented the phone, whereas Marconi wanted to improve the phone and invented the radio, he would have been outraged. Now, Marx was mistaken in many of his predictions: he predicted the advent of the dictatorship of the proletariat, said that work would eventually destroy humankind, and posited that exploitation would have reduced humans to dwarves. But his most erroneous prediction, in my opinion, was indeed the desirability of a multitasking life (which, incidentally, Marx conceives of as exclusively for males, as female multitasking already existed but for him, it seems, involved no intellectual activity).

Consumerism. Before the documedia worker there was the consumer, a figure typical of the media age. This period, located between that of the capital and that of documediality, is one where commodities are seen not from the standpoint of production but from that of consumption. Communication is largely destined to a function that in the manufacture age was very marginal: namely the promotion of consumption, or shopping tips. As soon as this evolution became apparent (and workers had access to the goods), people engaged in a criticism of consumption equal or superior, by force and indignation, to the much more motivated criticism of exploitation that had preceded it. If we reflect on the role of social objects and artifacts we understand how absurd and moralistic it is to criticize consumerism. Today this attitude is thankfully disappearing, only surviving in unlikely utopias like that of happy degrowth (which, to be consistent, should also advocate the decline of average life expectancy). However, this view has tormented my generation: it was as if you could only be human by giving up objects, whereas objects *are* what make us human, and spiritual. In its strong and immature moralism, the criticism of consumerism postulates the perfection of the human in the absence of objects, but the truth is that without objects, technologies, and documents the human would cease to be such.

Mobilization. I have mentioned the characteristics of the documedia workers (that is, the mobilized) when describing the inapplicability of the Marxist notion of "labor" to today's society. Now, it does not make sense to blame Marx for this. However, it does not make sense either to keep on taking the specter of capital as the cause of all evils, whereas what is actually happening is the global spread of the traits that Marx attributed to communism. As I mentioned earlier, the division of labor is disappearing, and so is the state (replaced by documedia instances operating on the Internet). Classes are disappearing too. (Do the bourgeoisie and the proletariat still exist? It is far from obvious.) Internationalism is being realized in the form of globalization, and the dictatorship of the proletariat has taken the form of populism. Are we happy now? Probably not, but this is not the point. One could say that the communist dream has come true in a way that wasn't expected or hoped for, but the gap between ideal and real is inevitable in the realization of any utopia, revealing its limits and clarifying its borders (even the most hyperbolic utopia, that of eternal life in Heaven, would pose very serious problems, starting with the boredom and depression that would likely come with it).

4   Recognition, Not Sustenance

This process seems to dispel the myth of the *Homo oeconomicus* and of the primarily economic motive of social interaction. Waste, which is usually interpreted as an exception or as a failure of economic strategies, is in fact ubiquitous in the documedia world, primarily in the sense of squandering time. However, one should make two clarifications in this regard. First of all, what is now coming to light is part of a deep and unanalyzed layer of social reality, which was there long before the documedia revolution. After all, even in the capitalist age, in 1914, masses of people went to the war full of enthusiasm. On the other hand, this is not simply a phenomenon linked to the introduction of negativity in the logic of profit. The fundamental motive of mobilization is not negation but self-affirmation, self-expression in public. Negativity, if it is there, turns out to be simply functional to this objective, which appears as recognition (à la Hegel). The struggle for recognition (which is often imaginary, as it mainly takes place with no antagonists in an echo chamber), once again, has no economic goals, or if it does have any, they are secondary: people want to be famous and then rich *in that order*, and no mobilized person would be happy to be wealthy but secretly so (Table 4).

Sustenance. Marx makes the same mistake that he attributes to capitalists when he isolates the reproduction of workforce as a kind of foundation or bare life, and this mistake, excusable then, is not forgivable today. Even in the age of manufacture there was probably more than such a robinson-

Table 4 From sustenance to recognition

Manufacture (production)	Sustenance
Mediality (communication)	Compensation
Documediality (recording)	Recognition

ade; for example the worker pride that Primo Levi describes in *The Wrench*. The book describes a very different situation from that of *If This Is A Man*, although Heidegger, in his timid condemnation of the Shoah, has compared the extermination camps to mechanized agriculture and, to be sure, to the atomic bomb. To think that sustenance is the sole end of the worker does not explain the fact that, once one has solved the sustenance problem, one can still continue to work in unpaid form, as is the case in documedia mobilization. In fact, today there is unpaid and unrecognized work whose evident counterpart is public recognition, whereas its hidden counterpart is the accumulation of data. Of course, Marx's descriptions of poverty do not correspond to anything that we know in the documedia world. They rather remind one of migrants, who represent a direct form of mobilization generated by war and need rather than documediality. But then, who are the mobilized of documediality? In terms of lifestyle, they are rather "comfortable," but they take the place of workers. They are not a class, but a function. Anyone can occasionally be mobilized, and the ontological persistence of classes has disappeared along with the persistence of activity or belonging. And what is the motive of the documedia worker, if not profit? Being intrinsically normative, documediality has a responsibilizing function, generating intentionality and moral anxiety: we are called to *answer to*, under the action of an appeal that is addressed to us individually and that is recorded (that is, it is addressed to us and cannot be ignored). However, responding passively is the source of *answering for*, responding in an active way, as bearers of morality and freedom: insofar as man is educated to the structure of answering-to he can later formulate the derived structure of answering-for (that is, being morally responsible). Obviously, those who transmit education can also do so intentionally (think of educators) or simply give the example without being aware of practicing a pedagogical function. What matters is that the intentionality and responsibility of those who receive education or witness the example are not a primitive, but a derivative, something that takes place a posteriori. Postulating a Kantian a priori responsibility, an *answering for* prior to the concrete experience of *answering to* means once again adopting the Pentecostal perspective, which is resolved in the assumption of a ghost limb called "intention," "understanding," "will." Hence, a fact I wish to underline. Contrary to a broadly shared view, the modernity or postmodernity in which the docu-media revolution takes place is therefore no more "liquid" and de-responsibilizing than the ages preceding it. On the contrary, it is the most granitic era of history, even when granite takes the apparently lighter form of silicon. In the documedia universe our intentions become as solid as trees or chairs: there is nothing more real than the Web, and this is where it derives its power from. Modernity has never been so solid. This means that, far from dissolving the structures of social reality, the documedia revolution has revealed them with unprecedented clarity and efficaciousness.

Compensation. In any case, in the stage after capitalism (that of mediality) there was already a huge gap. The thing at stake was no longer bare life, but what bare life is not: life as a human being. As with consumerism, it is absurd to consider this second stage as a worsening or even just a continuation of the previous condition: instead, it was a radical change from it and a decided step forward in the realization of the essence of human life. The latter, in fact, finds its raison d'être not in bare life, but in its supplements: from technology to culture, to culture as technology. From this point of view, the treatment undergone at the time by, say, the viewer of trash television is paradoxical: he was stigmatized as brain-dead, as a subcultural lumpen, and he was denied any form of the compassion abundantly given to the exploited worker. And yet there might very well be a same person who, after being exploited from 9 to 5, sits in front of the TV. Also, those who watch the worst reality shows still seem closer to human dignity than those working in an assembly chain: the latter just do what they are told, whereas the former at least exercise the modest freedom offered by the remote controller.

Recognition. This trend became stronger in the documedia age. The worker's annihilation mentioned by Marx was replaced by valorization and recognition, and the exploitation now takes place over the human capital embodied by the consumer (beauty, fashion, posts, cooking, mass vacations: of course one wonders if this can still be called exploitation). The main objective does not seem to be the economy, but rather the hope of recognition. Tolstoy noted that history is made by generals because they are the ones who write it, whereas if farmers wrote it the narrative would be very different. Well, this consideration also holds for society, which has been interpreted and described mainly by economists for the past three centuries. This, of course, cannot be taken as coincidental or arbitrary; however, choosing the economy as the ultimate element guiding every aspect of social reality leaves history full of holes and exceptions, like the ones listed by Bataille in his description of *dépense*: luxuries, mourning, wars, cults, monuments, games, shows, arts, non-reproductive sex. That these are no exception is proved by the fact that today they are mass phenomena (sex and reproduction are completely separate, for example, and art has turned out to be economically charged). This shows, on the other hand, that these are not situations that only belong to the past or to some foreign land, like human sacrifices or the gift of rivalry. If these things seem so exotic, it is only because we haven't thought enough about our world: in a way, the voluntary mobilized is not too different from the enthusiastic sacrificial victim in Aztec cults.

5   Self-Affirmation, Not Alienation

The hypothesis of recognition is the best one to account for the documedia mobilization. It may therefore be useful to break it down and highlight its constitutive elements. The old repetitive and alienating labor is replaced with a varied and revelational work. As I recalled above, the mobilized person is equal to the one described in Marx and Engels's *German Ideology*: a whole man who does everything. Sure, he is not happy, but that's a different story. Those who flock to the beaches or malls may be similar to the Fordist workers, but they want to do what they are doing with their whole heart; and they are better, they can be happy or dream of being so, whatever one means by the obscure term of "happiness." It would be an unforgivable mistake, and an injustice to their human dignity, to ignore this state of mind, and to bring it back to a form of false consciousness. The process is more complicated than that: neither productive alienation nor communicative distraction are purely and simply such, but they both reveal a will of self-affirmation. The latter, clearly shown by the documedia revolution, forces us to completely revise the category of "alienation." Indeed, alienation rests on a questionable (but dogmatic for its supporters) scale of values that follows a precise descending hierarchy: being, having, appearing. And "being," as in Heidegger, designates the lost, and purely imagined, homeland of authenticity.

Against this naturalistic or metaphysical idealization, postmodernists would praise appearance, which presents itself as the simple overturning of the hierarchy (what they call "overturning of Platonism," following Nietzsche). But this is not the point. The point is rather to reject the Pentecostal perspective with an emergentist perspective, considering being as something that is revealed thanks to technology, having, and appearing. In this sense, being is something that lies before us, not behind us (at the epistemological level). In the social world there is no being in itself independent of appearing and having; an event that no one has heard of is not simply unknown and therefore neither true nor false (as happens in the natural world): it rather does not exist. Alienation is considered an "aberration" (a clinical term dear to Marx and so many social critics), which obviously presupposes an otherwise positive compliance with the law of nature; something that is difficult not only to be found, but also to be conceived and hoped for. Do we really have so much nostalgia for a past that was machistic, unjust and (what's worse) boring? One should give up the myth of a natural humankind or of a humanity in itself, which only triggers nostalgic, catastrophistic, and recriminatory analyses that, most importantly, offer a pharisaic consolation. However, it is not clear why the transformations brought by capital and its successors (provided there have been any) should be interpreted as regression rather than as progress, and above all (this, I repeat, is the decisive aspect) as alienation rather than revelation (Table 5).

Alienation. "*It is now no longer the labourer that employs the means of production, but the means of production that employ the labourer,*" Marx emphatically wrote in *Capital* (the italics are his). This leads one to believe that there can be a non-alienated relationship between the human and technology, one where man is in control. And yet, any kind of relation-

Table 5 From alienation to self-affirmation

Manufacture (production)	Alienation
Mediality (communication)	Distraction
Documediality (recording)	Self-affirmation

ship with technology always entails a form of submission, which therefore depends on the nature of man and technology, not on capitalism. Now, every time you make a Nespresso you have to serve the machine as much as it serves you, without this being labor. But craftsmanship is also conditioned by (and finds its purpose in) the tool, and this also holds for intellectual performances: Euler said that his theories lay on the tip of his pen and Proust claimed he had no freedom with his books. Once more, Marx cultivated the illusion of a free craftsman who is alienated by a cruel world. But if Proust and Euler were not free, then who is? Now, I do not mean to compare young Werther to an English miner (Goethe was well aware of the differences, and so are we), but it seems to me that the relationship between human and technology should be interpreted as an essential form of revelation.

Distraction. Marx stigmatized the worker's alienation; Debord stigmatized the viewer's alienation in front of the observed thing. If (as I am trying to argue) the alienation of the former is problematic, then the latter is even more so. What's wrong with enjoying the spectacle? Is it not the contemplative state promoted by the Vedas and by Schopenhauer? Rather than alienation, it is distraction: you want to distract yourself from your thoughts and forget yourself. Once again, one wonders why self-oblivion produced by asceticism or action should be better than that produced by television. Indeed, Debord hates the society of spectacle almost as if it killed children as nineteenth-century capitalism did. This is a recurring attitude towards the media, which seem to have inherited (and deviated) the hatred one should reserve for more dangerous things, such as mass destruction weapons. On the one hand, a show is less gruesome than a battle, unless the show is by Hermann Nitsch. On the other hand, one wonders: distraction of whom, and from what? Consider those who are distracted, say, by a soap opera: would they otherwise be engaged with issues of infinitesimal calculus? They would be much more likely to join a pro-war demonstration, so they may as well be distracted.

Self-affirmation. From Plutarch to Hegel, thinkers have always been aware of the importance of recognition for human beings. But in the last few centuries humans have defined themselves in ways that were much too flattering, describing themselves as very (indeed, too) rational. Hence the myth of *Homo oeconomicus* and of instrumental rationality, which have filled up libraries and fueled learned debates. The documedia revelation disrupts these myths, manifesting the extremely uneconomic nature of our lives. Our goal is highly spiritual: self-affirmation in order to be recognized by the other. Why would anyone ever work for free, posting documedia content generating wealth only for the providers, buying one's own means of production (smartphone, laptop, and so forth)? Well, more than half the world does this, it seems: all those who use social networks. I personally do not, but that makes me no exception: I am still writing a book to affirm myself in the hope of being recognized by others. But the fact that professors are self-absorbed was well known; it wasn't known that everybody else is, too. From this point of view, nothing is more wrong (and, as often happens, moralistic) than interpreting selfies as a narcissistic act. Narcissus looked at himself and was content. Selfie-takers, instead, want to publish their pictures and do so not for their own pleasure but to be recognized by as many people as possible. The need for recognition is, at the same time, responsibility and mobilization, which is a source of normativity.

6   Atomization, Not Classification

I have not forgotten to mention post-truth. What I described so far is the technical and ideological framework that makes post-truth possible. Documents take the place of commodities or, more accurately, the mysterious nature of commodities lies in their intimate document-like nature; the needs of sustenance have diminished for much of the West, leaving more space to leisure time and therefore to the need for recognition (which is also there during work time). Even more radically, the distinction between work and free time has become very thin, meaning one should give up the notion of alienation and focus on notions of objective revelation and subjective self-affirmation. These transformations explain why truth (and mainly its subproducts and sophistications) has become the greatest production of the documedia age. "Moi, la Vérité, je parle," to quote a post-truthist *avant la lettre* like Jacques Lacan, who rightly attributed the need of self-affirmation to what today we call "post-truth." Self-affirmation becomes even more urgent now that the social fabric seems made up of monads: lots of individuals or micro-communities that constitute atomic wills to power. They represent themselves online neither by producing artifacts (as in the industrial economy) nor through the use of social objects (as in media society), but through the production of social objects (self-representations, statuses, selfies, contacts, e-mails and so on). Classes, as well as professional and social categories, are disappearing,[1] socially mirroring the phenomenon by which various activities like the production (and sale) of radios, televisions, telephones, calculators, computers, clocks, barometers, and so forth are summed up in the production and sale of smartphones. The monads have exclusive values shared in small groups, which in turn can communicate with other groups, but always horizontally, without the verticality (from an authority to the recipients) that characterized the industrial and media ages. Beliefs are atomized and privatized, defining one's identity: "I, Truth, post." This adaptation of Lacan's saying expresses the fundamental state of mind underlying post-truth. Let's have a closer look at the phenomenon of monadization (Table 6).

Classes. Classes play a very important role in Marx, especially with the related notions of "class struggle" and "class consciousness" (which are close relatives of the equally problematic notion of collective intentionality). However, the world wars, in which state apparatus and national affiliation were stronger than classes, showed how problematic this concept is. There is no such thing as collective intentionality, nor has there ever been. To win the war, Stalin had to reintroduce military grades and speak of a "great patriotic war." In short, where is the collective intentionality that should make us act as one and be the foundation of social reality? We share documents, and when these multiply, society is atomized. The emblem of this condition would be Mark Zuckerberg in a room where everyone is wearing Oculus, the virtual reality viewer. On the one hand, Zuckerberg here recalls Plato's philosopher, the one who is out of the cave while the others, chained, contemplate the simulations

Table 6 From classes to monads

Manufacture (production)	Classes
Mediality (communication)	Users
Documediality (recording)	Monads

projected inside. On the other hand, however, there are three significant differences. First, the captives here have voluntarily put on the mask, indeed, they bought it. Second, unlike Homer, he does not have to produce the contents of the show: the very people in chains are those who produce it by exchanging their own deeds on Facebook. Third and most important, he himself only has a privileged standpoint, some sort of class awareness that has nothing to do with that of Thomas Buddenbrook or even Felix Krull. If we do not consider this and continue to read the current situation with the lens of Marxist classes, our analyses are inevitably going to fail.

Users. As classes disappear, the only form of unity is found in the common taste of users in a mass society, in the world of hit parades and best-sellers. Except that, for some obscure reason, those who value class consciousness deplore the formation of something similar to a shared taste. Reading Debord (or, for that matter, and on the same basis, Adorno), one wonders what the desirable alternative to the society of the spectacle would be. A society without spectacles? A society of pedagogical spectacles? A society where spectacles are austere and only accessible to an elite like the Neue Musik proposed by Adorno? As always, it is hard to compare the present with the past. At one point, Debord deplores a decline in the quality of shows (without making any examples) when there are good reasons to argue that it has much improved. Indeed, a contemporary TV series has conceptual and technical subtleties that would have been unimaginable not only in commercial products 80 years ago, but also in Greek tragedy (provided of course that tragedy viewers were not themselves alienated, demented, and so forth). However, in the media phase there was a verticality to the system that is opposed to the horizontality of the documedia age. Debord insisted on separating the spectator from the performer, and considered this separation and division of labor as two essential characteristics of the spectacle. In this sense, divas would have the same origin as gods according to the later Feuerbach: pieces of consciousness of the spectator that are alienated and become superhuman and transcendent through the spectacle, so that, due to a specialization of labor, the show speaks in the name of the spectators. This transcendence of the media phase is the complete opposite of the documedia "one counts as one." However, even in this case the change has been progressive, and already in the media world there were processes announcing the documedia horizontality: quizzes, talk shows, reality shows, and all forms of trash TV. In some cases (talent shows) the narrative is that of the apotheosis: the passage from the human to the divine through the show.

Monads. Finally, let's come to monadization and horizontalization. The structures aimed at achieving goals (barracks, factories) of the production age, and the large audiences generated by communication, were followed by a documedia fragmentation creating social atomization (echo chambers, sense fields, and so on). Each monad reflects from its point of view (as a force representing the universe) everything that happens on the Web (the Library of Babel). Note well: the point of view is individual, implying a powerful anamorphosis of which post-truth is but one of the outcomes. In documediality, social divisions cease to matter: master and servant work in the same space and do the same things through the same media. At the same time, however, instead of massification there is a leveling, a horizontality of monads that takes the place of the capitalist class verticality. There are still some that are above the others, but there are so few of them that they are not a class: about a hundred people in the San Francisco area. What bourgeoisie could ever control systems such as the documedia ones? It's not like owning newspapers or televisions. Hence, the crisis of trust in the digital world (Domenicucci and Doueihi 2017). The modern ideological structures were capital, race, faith, homeland, conspiracy, and so on; that is, notions spread between wide and cohesive groups. Postmodern ideological structures, developed after the end of great narratives, represented a privatization or tribalization of truth (it is worth noting that the end of the great narratives coincides, consistently with the creation of "regional rationality," with the first cases of negationism) (Wieviorka 2017). Perfecting and spreading this process, documediality has created the atomism of millions of people who are convinced that they are right: not together (as believed mistakenly by the ideological churches of the past century) but on their own, or rather with the sole response of the Internet. The center of this new ideology lies in the pretense of being in the right regardless, and in seeking recognition through a technical apparatus: the Web. The latter facilitates the circulation of individual ideas while making them irrelevant (precisely because one counts as one) and realizes the microphysics of power, channeling all these ideas (different but all equivalent) into an immanent *like*, an apparatus that counts for the sheer numbers it expresses, manifesting a punctiform public opinion. This is me: that's the refrain underlying an inflation of self-made truths. Instead, in the age of ideologies, there was only one truth that was believed in by wide communities.

7   Conclusion

Have we reached the bottom, to use a moralistic (and optimistic) expression, always applied to all technological and social transformations? Of course not. Rather, based on the theory of revelation that I have tried to present in these pages, we are now closer to truth: social structures manifest themselves more clearly, and humankind reveals its nature. This, despite appearances, is progress (indeed, there is no argument to justify any talk of regression). More generally, every next stage reveals the essence of the previous ones: every stage shows dialectic advancement, not regression. Documediality reveals the decisive role it played in manufacture and mediality; the recorded act shows that both the commodity and the spectacle are indeed social objects; mobilization proves the overcoming of labor through free activity (qua non-retributed); the monads are the conciliation of class collectivity and user individuality in a single being representing the universe; the need for recognition and self-affirmation show human nature as the bearer of values; and revelation is the non-idealized manifestation of humanity. It is from here, from documediality, that one must start today to understand the ongoing social transformations as well as human nature, leaving capital to historians.

Note

1. This transformation is the focus of Richard Sennett's research. See in particular Sennet (1998).

References

Debord, Guy. 1994. *The Society of Spectacle*. New York: Zone Books.

Domenicucci, Jacopo, and Milad Doueihi, eds. 2017. *La confiance à l'ère numérique*. Paris: Éditions Berger-Levrault et Éditions Rue d'Ulm.

Ferraris, Maurizio. 2012. *Documentality. Why It is Necessary to Leave Traces*. New York: Fordham University Press.

Piketty, Thomas. 2014. *Capital in the Twenty-First Century*. Cambridge, MA: The MIT Press.

Searle, John Rogers. 1995. *The Construction of Social Reality*. New York: The Free Press.

Sennet, Richard. 1998. *The Corrosion of Character, The Personal Consequences of Work in the New Capitalism*. New York and London: Norton.

de Soto, Hernando. 2000. *The Mystery of Capital*. New York: Basic Books.

Wieviorka, Michel. 2017. Face à la "postvérité" et au "complotisme". *Socio* 8: 85–100.

Recording the Web

Janne Nielsen

1   Introduction

One of the most prominent media in the media landscape today is the world wide web (hereafter web).[1] A significant part of social, cultural and political life takes place online (at least in the West), and the web is a cultural resource offering information, entertainment and many ways to communicate and interact. The web is an ever changing scene character- ized by mutability and unpredictability as the technical infrastructure, the applications running on the web, and the content change and evolve rapidly. The ephemerality of the web is however accompanied by various recording practices and it is thus not only a place for communication but also for registration (Bordewijk and van Kaam 2002).[2] With the perva- sive use of technologies for tracking users and their behavior online, the web is central to commercial industry's comprehensive effort to record

J. Nielsen (✉)
Department of Media and Journalism Studies, Aarhus University, Aarhus, Denmark

© The Author(s) 2018
A. Romele, E. Terrone (eds.), *Towards a Philosophy of Digital Media*,
https://doi.org/10.1007/978-3-319-75759-9_4

and compile information about users. Tracking has been one of the driving forces behind the development of the web, influencing how the web looks and is experienced today.

The purpose of this chapter is threefold: first, to contribute to the theorizing about recording as a central feature of digital media by discussing whether or not the web is in itself a record, or rather a body of records to be compared with an archive. 'Recording' is used as a wider term for practices including registration.[3] Second, to study tracking as recording practices by describing different techniques for tracking, and to argue for the relevance of a historical view on tracking in order to understand how the landscape of tracking has changed over time. Third, to explore how another type of recording of the web, web archiving, is necessary for a comprehensive historical study of tracking.

Several studies of tracking exist but few offer a longer perspective on the spread and development of tracking techniques. A historical study of tracking could provide significant new knowledge about the role of the tracking practices (and the role of the powerful players behind the tracking) in shaping the web, while also shedding light on the social impact of tracking and the privacy issues involved. In order to examine tracking on the web in a historical context, we must first address the complexity and volatility of web. The ephemerality of the web means that the history of the web is not available to us online. Instead we must turn to web archives, but this raises some new challenges relating to the characteristics of archived web materials, and the question of what can be found in the web archives. We must understand how the archiving process affects what is archived, and how the nature of the archived web affects it as an object of study. With regard to the question of what can be found in the archive, we must explore whether trackers (or remnants or traces of tracking technologies) can be found in web archives. This requires a study of the complex landscape of tracking technologies to define what it is we are looking for, when we search the web archive for tracking technologies. In order to study the history of trackers, we must study these different types of recordings, and explore how one (tracking) is recorded in the other (web archives). The chapter concludes by addressing some foreseeable challenges in a study of tracking in web archives, thereby setting the frame for further studies of recordings on and of the web.

2   Is the Web an Archive?

The web is "the information source of first resort for millions of readers" (Lyman 2002), and we expect to be able to find information here about almost everything. The ready availability of an immense amount of information online can lead to an understanding of the web as a large library or archive with billions of records. In this light, it is relevant to examine whether we can in fact understand the web as an archive where everything happening online is recorded and information is stored indefinitely. It is true that a lot of information accumulates on the web, and some websites have archives built into their functionality. We often see this in, for instance, social media, news media and broadcasting media but also in people's personal websites or blogs. The notion of the web as an archive or library is supported by warnings and discussions about what information and material to publish online, and even more so what not to publish (ranging from indecent or libelous content to parents publishing pictures of their children without consent, and employers googling applicants or checking their Facebook and so on), and in the ongoing debate about the so-called "right to be forgotten" (see, for instance, the *Guardian*'s theme on the subject from 2015 to now). These discussions tend to strengthen the idea of the web as an archive, because they signal that there is a risk that something will be forever available and thus not forgotten. These are legitimate concerns, and the development of search engines definitely increases the risk of sensitive and/or possibly stigmatizing information about the individual being available online.[4] However, while it is true in principle that material published on the web can stay online indefinitely and thus can be found by people looking for it, if it is placed on a website that is maintained and indexed by search engines, it is more often than not the case that web content changes or disappears.

Media scholars are used to early historical material being scarce (when studying, for instance, radio and television history), but when it comes to web materials the problem tends to be more far-reaching than for other media. Brügger (forthcoming) describes the web as "volatile, subject to deletions or changes that may occur at an unprecedented scale and pace, compared to those in other known media types," and Ankerson notes that it is "far easier to find a film from 1924 than a website from 1994" (Ankerson 2012, 384). Contrary to common belief that the web is a big archive in itself, the nature of the web is in reality much more "fluid" (Day 2003, 1), as "pages or entire sites frequently change or disappear, often without leaving any trace" (ibid.). Following several studies on the lifespan of web pages and websites, it is, today, an established fact that content on the web is continually changed, moved and deleted at a fast pace (Cho and Garcia-Molina 2000; Lawrence et al. 2001; Lyman 2002; Agata et al. 2014; Jackson 2015; Kahle 2015). Across the studies, the average lifespan of a webpage is estimated to be between 44 and 100 days, and the percentage of content unrecognizable or gone after a year is around 50%.

The ephemerality of the web can be illustrated with the problem with link rot and reference rot. Link rot refers to the fact that a link can become invalid ("die" or break) in the sense that it points to a web resource (a website, webpage or element) which is no longer available, thus returning a 404 error, DNS error or similar "false result." Reference rot occurs when the link still works but the content has "decayed" in the sense that the information to which the link was supposed to point (the information referenced) has changed significantly or been replaced with something completely different. The problem with dead or broken links, which every web user has probably encountered, is becoming more serious with the increasing social, cultural and political role of the web, and as hyperlinks are used in more contexts. Several studies show that link rot and reference rot are causing significant problems in scholarly work (Lawrence et al. 2001; Koehler 2004; Day 2006; Klein et al. 2013; Massicotte and Botter 2017; Nyvang et al. 2017) and judicial cases (Liebler and June 2013; Zittrain et al. 2014) as short-lived links are used as references in academic articles and courtrooms.

The lack of persistence of web references underscores that although the web does hold billions of records, and some websites may have archival functions, the web in general is definitely not a library or archive. Articles, pictures, tweets, Facebook posts and so on can be deleted, often without leaving a trace. The nature of the web is rather "a unique mixture of the ephemeral and the permanent" (Schneider and Foot 2004, 115). While the web in itself does not record everything happening online, there are many players online interested in recording what happens online, specifically which users do what. The next section of this chapter will discuss practices of tracking on the web.

3   Tracking as a Way of Recording the Web

We're at the start of a revolution in the ways marketers and media intrude in (and shape) our lives. Every day, most if not all Americans who use the internet, along with hundreds of millions of other users from all over the planet, are being quietly peeked at, poked, analyzed, and tagged as they move through the online world. (Turow 2012, 1–2)

The web was originally conceived of as "a pool of human knowledge" (Berners-Lee et al. 2004, 907), a space for sharing information to the benefit of the users. But the web is not just an information space; it is also the largest advertising space available, and the interest in using the web to advertise for products and services, and in some cases to sell them online as well, has been one of the important driving forces in its development.[5] Advertising has played a major role in how the web looks today, and in the shaping of some of the most prominent players online like Google and Facebook. The invention of technologies for storing and tracking information about the users has been pivotal in this development, creating an online environment where the user is considered a costumer, and where companies follow the users' movements around the web to learn as much as possible about them. This knowledge is then used to present users with targeted information in the hope that it will lead to purchases (or clicks, depending on the business model). The collection and sale of information about user characteristics and behavior online is a huge industry, but it also raises serious concerns about the privacy of the users, especially since the tools for recording and registering user information and behavior are now prolific and intrusive in a way that cannot be compared with the registration capabilities of previous media. The following is not an attempt to write the history of advertising on the web,[6] nor to give a comprehensive account of the development of tracking online. The aim is to present tracking as a way of recording the web, and discuss the implications of the different tracking technologies as a first step in further studies of how the history of tracking can be written by using web archives.

The first clickable ad was sold in 1993 by Global Network Navigator (Simpkins et al. 2015). The following year the first banner ad was placed on a website by HotWired (D'Angelo 2009), and the cookie was invented at Netscape Communications along with cookie-placement capability integrated in the Netscape Navigator browser (Turow 2012, 47). This set the scene for the spread of advertising online and for ad-serving companies like DoubleClick to enter the market.[7] Although it took some years before online advertising really caught on (at first companies were not convinced of the value of online ads), it marked the beginning of what was to become a huge market for advertising online and a significant driving force for the development of the web.

The use of technologies for tracking has, of course, changed over the years, but the HTTP cookie (hereafter cookie), the first technique used for tracking, is probably still the most well-known type of tracker. Many web users are familiar with the term 'cookie' as they will likely encounter it regularly on websites they visit.[8] The cookie is technically a small text file with an identification number, which is placed on the user's computer. The information does not generally identify the user but the user's browser; so, using a different computer or browser (or deleting all cookies in the browser) will mean that the user will appear to the website as a new user (until a new cookie is set). The purpose of cookies is to hold information about the user's interaction with a website.[9] This is commonly used to remember a user's credentials, preferences or choices; for example, logins, choice of language, volume of a player, or items added to a shopping basket. This can be used to personalize content and settings on a website, and it can save the user re-entering information on successive webpages or later visits to the website. In this way, cookies can help to improve the user experience. A cookie can also contain information about the clicks made by the user while visiting a website, and this information can then be accessed by the web server of the website in the case of repeat visits. It is this information about preferences and behavior which makes the cookie a useful tool for advertising.

There are different types of cookies, and some are generally more benign than others. Cookies can be classified by lifespan and domain as either a *session cookie* or a *persistent cookie* depending on whether they are erased after the session (that is, when the user closes the browser) or remains on the device for a set period of time; and either a *first-party cookie* set by the server of the visited website (domain) or a *third-party cookie*, where the cookie is included or embedded in a site different from the owner of the cookie (outside the domain of the visited website) (EU

Internet Handbook). Third-party permanent cookies are potentially more "hostile" as the information is not just confined to the website visited, and the information can be stored over a long time.

Cookies are a way of recording web practices, in that a cookie might register the above-mentioned information about the user. It can record where a user is when, and if you click an ad, the when and where of this is recorded along with information about, for instance, the country you were connecting from. One website can set many cookies. In 2009 a study commissioned by the *Wall Street Journal* found that the 50 most-visited US websites left an average of 64 trackers (HTTP cookies, Flash cookies, and beacons) on users' computers, with one website leaving as many as 234 trackers (Angwin 2010). A 2015 study by Altaweel et al. (2015) found 128 HTTP cookies in average on the top 100 websites as well as other trackers. This speaks to the magnitude of the recordings taking place.

The limitation of HTTP cookies from an advertiser's or an ad-serving company's point of view is that users can with relative ease delete cookies, and as the technology has become more known and the privacy issues debated, more users delete their cookies on a regular basis (Soltani et al. 2009; Weinberger 2011). Online advertising companies have an ongoing, strong incentive for finding more reliable tracking methods, resulting in the development of new tracking technologies. One option is to take advantage of the local storage in Adobe Flash to store Local Shared Objects, also called Flash cookies. Contrary to a common cookie, Flash cookies are not connected to or controlled by a browser, so different browsers can access the same Flash cookies, and Flash cookies are not removed by clearing cookies through the browser's settings. Even with private browsing or incognito-modes of browsing, enabled Flash cookies can still work and track the user (Soltani et al. 2009). Flash cookies can also store much more information than HTTP cookies, and they do not have a default expiration date (ibid.). Some studies have shown between 20% and 50% prevalence of Flash cookies on top websites (Soltani et al. 2009; McDonald and Cranor 2011; Ayenson et al. 2011). HTML5 Local Storage is another client-side storage mechanism used for tracking, shown to be prevalent on 76 of the top 100 websites (Altaweel et al. 2015). In 2012 most of the web tracking was, according to Roesner et al. (2012), done using cookies and local storage mechanisms such as Flash cookies and HTML5 local storage.

With the diffusion and development of tracking technologies, new practices have evolved with the aim of counteracting the efforts of users to delete trackers. By applying multiple tracking technologies, the risk that the user finds and deletes all types of cookies is diminished. Several studies show how Flash cookies can even be used to reinstate other cookies deleted by the user, a process called "respawning" (Soltani et al. 2009), thereby circumventing the user's attempts to avoid tracking (Soltani et al. 2009; Angwin 2010; Ayenson et al. 2011; Acar et al. 2014). ETags identifying specific resources at a URL can also be used for respawning,[10] which, according to Ayenson et al. (2011, 11), is "particularly problematic because the technique generates unique tracking values even where the consumer blocks HTTP, Flash, and HTML5 cookies". Additionally, private or incognito browsing mode does not prevent tracking by ETags, and the only way to block the tracking is to always clear the cache after visiting a website (ibid.). Cookies can also be combined with another tracking device called a web bug (or beacon), which is a bit of software code, typically a 1-by-1 pixel image (Maclay 2009), embedded in a webpage, which can record visits to a website and transmit data about a user's behavior on a site (e.g. clicks, mouse movements or typing) (Angwin 2010).

Ad-serving companies and ad networks can have their cookies embedded on a multitude of websites (with different domains), which allows them to compile information recorded during the use of all these websites, and to track users across websites. They can also link cookies in a process called "cookie synching", where different trackers share their user identifiers in order to exchange user data across multiple platforms (Tene and Polonetsky 2011, 291; Acar et al. 2014).[11] The respawning of cookies can allow for further synching of identifiers across domains (Ayenson et al. 2011). By using different storage mechanisms, deleted cookies can even be restored by other cookies or other tracking technologies, creating "evercookies" (Ayenson et al. 2011; Acar et al. 2014) or "zombiecookies" (Mayer and Mitchell 2012). One evercookie uses up to 17 different storage mechanisms, where available, to create an extremely persistent cookie (Kamkar 2010).

Other advances in tracking technologies include different ways of creating fingerprints of browsers (or other devices). A technique called "canvas fingerprinting" (Mowery and Shacham 2012) exploits the HTML5 canvas element to render an image that identifies the user's browser because the same image is rendered differently depending on operating system, browser, settings and so on (Eckersley 2010; Mowery and Shacham 2012; Shen 2014). Some types of fingerprinting require a script or plug-in, which actively discovers certain properties of the browser, like fonts, enabled plugins, enabled first-party and third-party cookies and so on (Mayer and Mitchell 2012, 421), while others use properties already available in the network traffic sent from the browser, like IP address, operating system, HTTP accept headers and so forth.[12]

The availability of various tracking techniques and the ways to combine them mean that the web hums with recording activities under the surface. Different trackers record varying information and properties. Some recordings are analyzed right away in order to present the user with expected relevant content in real time as the user moves around the web. The recordings are also accumulated in databases, where they can be combined with other recordings pertaining to the same user ID allowing for ad networks to create comprehensive profiles of users, which can be used to draw conclusions about what the user might be interested in and to group users in segments or target groups based on their assumed demographics, interests, economic capabilities and so on. These records, potentially containing personal and private information about users, can then be sold to advertisers so they can target their ads at expected relevant users. The use of registration and tracking for advertising purposes is cause for concern, especially because of the privacy issues involved. A lot of this information is in the hands of a relatively small number of very large and powerful companies (Krishnamurthy and Wills 2009). A case in point is that studies have found Google tracking technologies to be present on 92%–97% of the top 100 U.S. websites (Ayenson et al. 2011; Altaweel et al. 2015).

Much of the information is collected and shared without the user's knowledge in what Turow (2012, 1) calls "one of history's most massive stealth efforts in social profiling". The invisibility of many of these tracking technologies is particularly concerning because it means that many users are not aware of the extent to which the tracking takes place. In the case of the cookie, paradoxically, the familiarity due to the overall presence of cookies might contribute to an underestimation of the possible impact and severity of the ubiquitous tracking devices. In addition to the privacy issues, the problem is, as described by Turow (2012, 2), that the profiling resulting from the analysis of user data leads to social discrimination as different users are exposed to different types of content depending on their profiles. When profiling is used to determine who is a valuable target for advertising, and who is not, much of what we are exposed to on the web will be a result of one or more labels assigned to us by marketers.

There is an interplay between tracking and the generation of specific content on a dynamic website. The recording is used for personalization, which might be to the advantage of the user because it eases the use and supposedly heightens the relevance of the content the user is exposed to, which supports the often heard claim that tracking devices are meant to help enhance users' experiences. But the same technology allows for a massive compilation of data, which the user would probably in many cases find unacceptable if they knew how much is actually saved and sold! This dual nature was inherent even in the first tracking technology. The cookie was not invented as a tracking tools in the negative sense but for usability reasons (Tene and Polonetsky 2011; Turow 2012). According to the inventor, Lui Montulli from Netscape Communications: "The goal was to create a session identifier and general 'memory' mechanism for websites that didn't allow for cross site tracking" (Montulli 2013). To reduce the potential of the cookie as "a universal tracking mechanism" (Turow 2012, 55), Montulli designed limits for the information sent back to the website and for who could read or change a cookie, the so-called *same-origin policy* (see Note 11); but, as shown above, workarounds inevitably followed. Regardless, the decision to make the cookie work across websites owned by the creator without permission of the user charted the course for tracking technologies as hidden objects, disguising their presence and significance from the user.

Existing studies of tracking speak to the prevalence of specific tracking technologies at specific times, but most studies of tracking report on a relatively limited amount of time. A historical study of tracking would contribute with a longer perspective on the development and diffusion of tracking technologies, hopefully revealing the rise and decline of different trackers, fluctuations in use and various combinations of technologies. It would also facilitate more valid comparisons over time, which is challenging without a longer historical view because it is difficult to judge whether differences in results are caused by actual changes in the object of study or by differences in methods (several of the studies mentioned in this chapter specifically point to this issue). In order to study the history of tracking, we must turn to another recording practice on the web: web archiving.

4 Web Archives as a Way of Recording the Web

If the web in itself were an archive, we would be able to study its history just by looking at it. But, since the web is characterized not just by durability but also volatility, the web of the past is, to a large extent, no longer available online. The realization that the web is an important part of culture and society, and that part of this history is relentlessly disappearing, has led to the initiation of several web-archiving initiatives.[13] Web archiving can be understood as a way of recording the web in the sense that it is a practice that attempts to register and deposit a copy of web data in an archive. The International Internet Preservation Consortium (IIPC), a member organization for web-archiving institutions, offers this definition: "Web archiving is the process of gathering up data that has been published on the World Wide Web, storing it, ensuring the data is preserved in an archive, and making the collected data available for future research."

Most web archiving archives not only content but also structure. Large-scale web archiving is most commonly undertaken by using web crawlers (Thomas et al. 2010); that is, software applications that move systematically around the web and collect as many web objects as possible. A crawler harvests content from an assigned list of URLs, but it also harvests links from these URLs and (depending on the settings) follows the links in order to archive the content linked to (cf. Nielsen 2016 for an introduction to web archives and web archiving, including different archiving strategies).[14] The archived objects can be replayed in the Wayback Machine, a software which can be used to view archived webpages in a browser and "surfing" the archive by following links in a similar way to the experience of the online web. It is also possible to do quantitative analysis of objects in the archive by accessing, processing and analyzing, for instance, index files or metadata files (for an example, see Brügger et al. forthcoming).

The web-archiving institutions work under different conditions and laws but in the cases where the purpose of a national web archive is to preserve the nation's online cultural heritage, the web archiving can be described as the state's (or more specifically a state institution with a legal mandate) way of recording what takes place in the national web sphere. In the case of Denmark, the national Danish web archive Netarkivet harvests and preserves the Danish web sphere (see Brügger et al. 2017), consisting of all domains on the Danish Top Level Domain (TLD) .dk and all other domains that are considered relevant for a Danish context and Danish culture (Schostag and Fønss-Jørgensen 2012). This archiving has taken place since 2005, after the dynamic internet became subject to the Danish legal deposit law in 2004 (Act on Legal Deposit of Published Material 2004). With this law, the Danish web sphere is now considered part of the Danish cultural heritage, and the national web archiving initiative is the country's way of registering, collecting and preserving for posterity web documents and other web content, which are part of the national public sphere.

The recording of what was on the web is important for several reasons, including the preservation of cultural heritage and as a source for future historians and other scholars studying any subject, which in one way or another has a presence online, but also as a way of documenting matters of public interests. For instance, when Malaysia Airlines Flight 17 crashed in 2014, the Internet Archive captured a webpage where the Ukrainian separatist Igor Girkin boasted about shooting down a plane (Dewey 2014). The comment was later erased but by then it was already in the archive available for all to see. In 2010 the Conservative Party in UK removed a backlog of speeches from its website (Ballard 2013) but the files had already been archived by the UK Web Archive and the Internet Archive (Webster 2013).

From a critical point of view, the recording of the web by cultural heritage institutions offers a way for the state to accumulate massive amounts of information about its citizens. The national Danish web archive Netarkivet, for instance, archives more than 1 million websites per year, many of them several times in a year (Netarkivet). This might include content which was only supposed to be online for a short amount of time and not meant to be preserved, and content which was not as such intended for the public sphere even though it was online (for example, people's private websites). The amount of data allows for big-data analysis, linking previously discrete information in ways that were not possible in previous times, which points to some of the ethical challenges relating to big data. In the case of Denmark, the web archive is only accessible in relation to research use, and the data protection law is very strict so as to prevent abuse of the material. The US non-profit organization the Internet Archive, however, makes its web archive of more than 510 billion web objects (Goel 2016) available to the public online.[15] The accessibility makes it an immensely valuable source material, while also increasing the risk of misuse.

Conceptualizing web archiving as a way of recording the web is, however, not without pitfalls. The ambiguity of the web as a complex interplay of the ephemeral and the permanent is mirrored in web archives, making the archived web an even more complex object of study than the live web. The main problem is that it is not possible to archive everything. One of the reasons for this is that a lot of content is continuously added, updated or removed, so to archive everything requires a constant effort. But even with great efforts, the archiving itself takes time, which means that websites may be updated while they are being archived. Brügger calls this "the dynamic of updating" (Brügger 2005), explaining how we can never know if, where and when a website might have changed during the archiving process. This has substantial implications for how we can understand what we end up with as an archived version of a website. We cannot know how close what we have in the archive resembles what was online. We cannot know whether we have lost something in the process and what we have lost; and if we archive websites with a high update frequency, we can be almost certain that we have lost something! But another aspect of the dynamics of updating is, arguably, even more serious, as Brügger points out. During the time it takes the archiving software to move through a website and archive it, some objects might have changed, thus the objects no longer have the exact same temporality. In this way, an archived version of a website may combine elements that were never online at the same point in time:

> Not only do we lose something that was there, we are also in danger of getting something that in a way was never there—something that is different from what was really there. … We thus face the following paradox: on the one hand, the archive is not exactly as the website really was in the past (we have lost something), but on the other, the archive may be exactly as the Internet never was in the past (we get something different) (Brügger 2005, 23).

What is available in web archives thus comprises not records of the online web but rather versions or reconstructions, which only resemble what was once online.

The challenges of recording the web (and of using the archived web as a source) stem not just from the temporal issues related to updating and archiving, but also from other aspects of the complex and dynamic nature of the web. Today, many websites are dynamic, containing client-side and/or server-side scripting (for instance, JavaScript and other programming languages) to generate content on the fly. This means that the website does not exist as a static object for the user to visit but instead is built in the browser, when the user accesses it, adding different elements to the site, which can, for instance, be dependent on the geolocation of the user, the date or time of day, the user's previous behavior online, and other factors. This is a big challenge for web-archiving software. Likewise, dynamic content involving JavaScript, Flash or interactive social media is notoriously difficult to archive (Schostag and Fønss-Jørgensen 2012, 120). The web links many different document types and content structures, which might require different archiving strategies (Day 2006), and the process of archiving entails several choices with regards to software and configurations that will influence the object (Masanes 2005, 77). It is also important to stress that a large part of what is online is not archived because the harvesting software cannot gain access. Some estimate that as much as 90% of all content on the web is inaccessible (Dougherty et al. 2010, 8) because it is stored behind forms-based query interfaces or is in other ways part of the so-called deep web or hidden web, which is not indexed by search engines. For instance, objects in databases or on FTP servers, or any content on websites with access restrictions (for example, password protection, requirements for IP authentication, or "nontrivial interaction" [Masanes 2005] such as CAPTCHA)[16] is not collected (for a longer discussion see Nielsen 2016).

It is, therefore, important to point out that recording the web does not mean making a copy of something, and that the web is not reproducible in the way other media are. If you record a radio program, the recording will be similar, if not identical, to the recording. This is not the case when the web is recorded. The archiving of a web object can better be described as the process of creating a new object on the basis of what is online. Pointing to this process, Brügger defines the archived web as "re-born digital" material: "Reborn digital material is born-digital material that has been collected and preserved, and that has been changed in this process to such an extent that it is not identical to the born-digital material from which it was made" (Brügger forthcoming).

In the case of the web, the record will seldom (if ever) be an exact copy of what was online. Consequently, in order to use web archives for a historical study of tracking, it will be necessary not just to understand the general challenges of using web archives but also to explore the specific challenges relating to how the trackers have been recorded and preserved in the archive.

5   Conclusion: Towards a Historical Study of Tracking

Although the web is not a record in itself, recording of the web is inherent in some of the technologies inhabiting the web. As shown, the diverse tools for tracking use different strategies for learning more about the users of the web, and the recordings of information and behavior are continually shaping the web and the experiences of the web. The widespread use of registration and tracking, and the issues arising from this, necessitate a scholarly focus on these practices. In order to take the first steps towards examining the rise and development of tracking practices on the web, this chapter has addressed the complexity and volatility of the web, the challenges relating to web archiving and the characteristics of archived web, and the diverse tracking technologies in use on the web.

Several studies describe the workings and prevalence of trackers, but studying tracking technologies in web archives and not on the live web will require new methods. Some of the existing studies of tracking are historical in the sense that they include longitudinal data; but only one study, to my knowledge, includes materials from web archives: Anne Helmond's case study of the historical development of trackers on the website of the *New York Times* (Helmond 2017). In this study Helmond shows how the traces of trackers and similar web objects can be found in the source code of archived websites (Helmond 2017, 145).[17] Helmond's work and methodology are very inspiring, but her method may not be applicable on web archives that are not accessible in the same way as the Internet Archive.

To study the evolution of tracking through web archives, it is necessary to investigate to what extent different tracking technologies (or remnants or traces of tracking technologies) can be found in web archives. How have they been influenced or changed by the archiving process? What tools can be used to access the code of the objects in the archive? Can Helmond's method (Helmond 2017) be applied or adapted to other archives than the Internet Archive? Could other methods from studies of tracking on the live web be repurposed for studies in web archives? Does the environment (the infrastructure) of the web archive influence the code of the trackers? In the case of HTTP cookies, for instance, the cookies as such will not be in the archive but the code that sets the cookie, that is the instruction from the server to the browser about storing information (and sending it back), has likely been recorded and preserved. But we need a tool that can access and extract this information in order to analyze it, and the tool has to be fitted to the infrastructure of the archive. Cookies that rely on, for instance, JavaScript may be a challenge, because some JavaScript is difficult to archive (Schostag and

Fønss-Jørgensen 2012) and caused problems especially in the early years of web archiving. The technology used to collect and preserve web objects has, of course, changed over time. A historical study will, therefore, also have to consider in what ways the results might have been skewed by technical differences in what could be archived at different times. Questions and considerations like these are essential in order to develop applicable methods for analyzing the historical development of tracking technologies as they are represented in web archives. This will hopefully help us understand how tracking has evolved over time, the development and diffusion of different tracking technologies, and the power relations behind the companies involved in the tracking. It might also help shed light on the everyday experiences of users in a setting where part of what you see it a result of someone recording, analyzing and responding to your previous and current behavior.

Notes

1. The two terms the web and the internet are often used interchangeably but they are not the same. The internet is a global network of computer networks, while the web is a software system built on top of the internet, using the Hypertext Transfer Protocol (HTTP), Uniform Resource Locators (or Uniform Resource Identifiers [URIs]) and Hypertext Mark-up Language (HTML) to communicate (see Berners-Lee et al. 2004 for a description of the architecture of the web). The web is "an information space" (Berners-Lee et al. 2004) on the internet, a way to access information on the internet, but the internet has many other functions and carries a lot of information not available through web browsers. This chapter focuses on the web and not the entire internet.

2. In Bordewijk and van Kaam's (2002) classical categorization of information traffic patterns, they define registration as "the issue of information by an information service consumer under the programmatic control of an information service centre" (Bordewijk and van Kaam 2002). Registration is thus characterized by an information center collecting information from a consumer about the consumer and his or her behavior without the consumer controlling what is collected.

3. To record can mean, inter alia, "to deposit an authentic official copy of" (Merriam-Webster) or to cause data to be registered on something "in reproducible form". To register can mean, inter alia, "to make a record of," to make "official entry in a register" or to "record automatically."

4. This issue is amplified with the new tools for big-data analysis of aggregated data.

5. The early internet, before the web, had guidelines that outlawed any commerce, including marketing (McCullough 2014).

6. For an account of web advertising history, see Crain (forthcoming).

7. DoubleClick was one of the first Application Service providers in internet ad serving, established in 1996 and acquired by Google in 2007 for $3.1 billion (Simpkins et al. 2015).

8. At least in Europe, where a European Union Directive from 2011, popularly named the "EU cookie law" (Stringer 2012), requires website owners to inform users about the use of most cookies and to get consent from users to store and retrieve information by using cookies (European Commission).

9. HTTP is a stateless protocol, which means that user data is not persistent between webpages. Without cookies (or other tokens or identifiers), a server has no way of knowing if two requests are coming from the same browser (and hence user).

10. When visiting a website, ETags are cached by the browser so the web server does not have to send a full response on subsequent visits if the resources are unchanged.

11. Cookies are domain-specific, meaning that they can only be read or changed by the domain that created them, the so-called *same-origin policy*. Cookie synching is a way of circumventing this principle, while also enabling "back-end server-to-server data merges hidden from public view" (Acar et al. 2014).

12. For more on fingerprinting, see, for instance, Eckersley (2010), Nikiforakis et al. (2013), Acar et al. (2014).

13. This chapter focuses on web archiving performed by cultural heritage institutions (so-called "macro-archiving" ([Brügger 2005, 2013]), while recognizing that web archiving also happens in other forums and contexts (see Brügger 2011).

14. Similar crawlers are a used for other automated tasks online, including updating and indexing.

15. Internet Archive collects about a billion pages every week (Kahle 2015). It also has many other digital collections, as it aims "to provide Universal Access to All Knowledge." (Internet Archive).

16. CAPTCHA (Completely Automated Public Turing test to tell Computers and Humans Apart) is a test used to ensure users are human by asking them to perform an interpretative task, typically recognizing and typing in twisted letters, which would be illegible for a computer.

17. To detect the trackers, Helmond and colleagues have created the tool *Tracker Tracker* by repurposing Ghostery, a browser extension for detecting and blocking tracking technologies (Ghostery). The *Tracker Tracker* tool on top of Ghostery makes it possible not only to detect and block trackers on each website visited but to create "*a network view* of websites and their trackers" (Helmond 2017, 146, emphasis in original), thus shedding light on the relationships between trackers.

References

Acar, Gunes, Christian Eubank, Steven Englehardt, Marc Juárez, Arvind Narayanan, and Claudia Díaz. 2014. The Web Never Forgets—Persistent Tracking Mechanisms in the Wild. In CCS' 14. *ACM Conference on Computer and Communications Security*, Scottsdale, AZ, November 3–7, 2014.

Act on Legal Deposit of Published Material. 2004. Translation of Act No. 1439 of 22 December 2004. Accessed September 9, 2017. http://www.kb.dk/en/kb/service/pligtaflevering-ISSN/lov.html. Archived version available in Internet Archive: http://web.archive.org/web/20170715184105/http://www.kb.dk/en/kb/service/pligtaflevering-ISSN/lov.html

Agata, Teru, Yosuke Miyata, Emi Ishita, Atsushi Ikeuchi, and Shuichi Ueda. 2014. Life Span of Web Pages: A Survey of 10 Million Pages Collected in 2001. In *JCDL'14. Proceedings of the 14th ACM/IEEE-CS Joint Conference on Digital Libraries*, London, September 8–12, 2014, 463–464.

Altaweel, Ibrahim, Nathaniel Good, and Chris Jay Hoofnagle. 2015. Web Privacy Census. *Technology Science*, December 15, 2015. https://techscience.org/a/2015121502/

Angwin, Julia. 2010. The Web's New Gold Mine: Your Secrets. *The Wall Street Journal*, July 31, 2010.

Ankerson, Megan Sapnar. 2012. Writing Web Histories with an Eye on the Analog Past. *New Media & Society* 14 (3): 384–400.

Ayenson, Mika D, Dietrich J Wambach, Ashkan Soltani, Nathaniel Good, and Chris Jay Hoofnagle. 2011. Flash Cookies and Privacy Ii: Now with Html5 and Etag Respawning. *SSRN*, July 30, 2011. https://papers.ssrn.com/sol3/papers.cfm?abstract_id=1898390

Ballard, Mark. 2013. Conservatives Erase Internet History. *ComputerWeekly.com*, November 12, 2013. http://www.computerweekly.com/blog/Public-Sector-IT/Conservatives-erase-Internet-history. Archived version available in Internet Archive: http://web.archive.org/web/20170822090538/http://www.computerweekly.com/blog/Public-Sector-IT/Conservatives-erase-Internet-history

Berners-Lee, Tim, Tim Bray, Dan Connolly, Paul Cotton, Roy Fielding, Mario Jeckle, Chris Lilley, et al. 2004. "Architecture of the World Wide Web, Volume One." Edited by Ian Jacobs and Norman Walsh. W3.org. December 15. https://www.w3.org/TR/webarch/. Accessed 28.09.2017. Archived version available in Internet Archive: http://web.archive.org/web/20170922014108/http://www.w3.org/TR/webarch/

Bordewijk, Jan L., and Ben van Kaam. 2002 [1986]. Towards a New Classification of Tele-Information Services. In *McQuail's Reader in Mass Communication Theory*, ed. Denis McQuail, 113–124. London: Sage.

Brügger, Niels. 2005. *Archiving Websites*. Aarhus: The Centre for Internet Research.

———. 2011. Web Archiving. Between Past, Present, and Future. In *The Handbook of Internet Studies*, ed. Mia Consalvo and Charles Ess, 24–42. Chichester: Wiley.

———. 2013. Web Historiography and Internet Studies. Challenges and Perspectives. *New Media & Society* 15 (5): 752–764.

———. forthcoming. *The Archived Web: Doing History in the Digital Age*. Cambridge, MA: MIT Press.

Brügger, Niels, Ditte Laursen, and Janne Nielsen. 2017. Exploring the Domain Names of the Danish Web. In *The Web as History*, ed. Niels Brügger and Ralph Schroeder, 62–80. London: UCL Press.

———. forthcoming. Methodological Reflections About Establishing a Corpus of the Archived Web: The Case of the Danish Web From 2005 to 2015. In *The Historical Web and Digital Humanities. The Case of National Web Domains*, ed. Niels Brügger and Ditte Laursen. London: Routledge.

Cho, Junghoo, and Hector Garcia-Molina. 2000. The Evolution of the Web and Implications for an Incremental Crawler. In *VLDB '00. Proceedings of the 26th International Conference on Very Large Data Bases*, 200–209. September 10–14, 2000.

Crain, Matthew. forthcoming. A Critical Political Economy of Web Advertising History. In *The Sage Handbook of Web History*, ed. Niels Brügger and Ian Milligan. London: Sage.

D'Angelo, Frank. 2009. Happy Birthday, Digital Advertising! *AdAge*, October 26, 2009. http://adage.com/article/digitalnext/happy-birthday-digital-advertising/139964/. Archived version available in Internet Archive: http://web.archive.org/web/20170731095446/http://adage.com/article/digitalnext/happy-birthday-digital-advertising/139964/

Day, Michael. 2003. Collecting and Preserving the World Wide Web. *Citeseerx. Ist.Psu.Edu*. JISC & The Wellcome Trust.

———. 2006. The Long-Term Preservation of Web Content. In *Web Archiving*, ed. Julien Masanes, 177–199. London: Springer.

Dewey, Caitlin. 2014. How Web Archivists and Other Digital Sleuths Are Unraveling the Mystery of MH17. *The Washington Post*, July 21, 2014. https://www.washingtonpost.com/news/the-intersect/wp/2014/07/21/how-web-archivists-and-other-digital-sleuths-are-unraveling-the-mystery-of-mh17/?utm_term=.ba61df3f8a1c. Archived version available in Internet Archive: http://web.archive.org/web/20160702071156/https://www.washingtonpost.com/news/the-intersect/wp/2014/07/21/how-web-archivists-and-other-digital-sleuths-are-unraveling-the-mystery-of-mh17/

Dougherty, Meghan, Eric T. Meyer, Christine McCarthy Madsen, Charles Van den Heuvel, Arthur Thomas, and Sally Wyatt. 2010. Researcher Engagement with Web Archives: State of the Art. *JISC Report*.

Eckersley, Peter. 2010. How Unique is Your Web Browser? In *Privacy Enhancing Technologies. Proceedings from 10th International Symposium, PETS 2010*, ed. Mikhail J. Atallah and Nicolas J. Hopper, 1–18. London: Springer.

European Commission. n.d. EU Internet Handbook: Cookies. Accessed September 14, 2017. http://ec.europa.eu/ipg/basics/legal/cookies/index_en.htm. Archived version available in Internet Archive: http://web.archive.org/web/20170902160903/http://ec.europa.eu/ipg/basics/legal/cookies/index_en.htm

Ghostery. n.d. About Ghostery. Accessed September 11, 2017. https://www.ghostery.com/about-ghostery/. Archived version available in Internet Archive: https://web.archive.org/web/20170901155310/https://www.ghostery.com/about-ghostery/

Goel, Vinay. 2016. Defining Web Pages, Web Sites and Web Captures. *Blog.Archive.org*, October 23, 2016. https://blog.archive.org/2016/10/23/defining-web-pages-web-sites-and-web-captures/. Archived version available in Internet Archive: https://web.archive.org/web/20171011052636/https://blog.archive.org/2016/10/23/defining-web-pages-web-sites-and-web-captures/

Helmond, Anne. 2017. Historical Website Ecology: Analyzing Past States of the Web Using Archived Source Code. In *Web 25: Histories From the First 25 Years of the World Wide Web*, ed. Niels Brügger, 139–155. New York: Peter Lang.

IIPC. n.d. About. Accessed September 19, 2017. https://netpreserveblog.wordpress.com/about/. Archived version available in Internet Archive: https://web.archive.org/web/20170722204633/https://netpreserveblog.wordpress.com/about/

Internet Archive. n.d. About the Internet Archive. Accessed September 19, 2017. https://archive.org/about/. Archived version available in Internet Archive: https://web.archive.org/web/20170715085801/http://archive.org/about/

Jackson, Andy. 2015. Ten Years of the UK Web Archive: What Have We Saved? Presentation from the 2015 IIPC General Assembly, Palo Alto.

Kahle, Brewster. 2015. Locking the Web Open, a Call for a Distributed Web. *Blog.Archive.org*, February 11, 2015. http://blog.archive.org/2015/02/11/locking-the-web-open-a-call-for-a-distributed-web/. Archived version available in Internet Archive: https://web.archive.org/web/20150305064916/https://blog.archive.org/2015/02/11/locking-the-web-open-a-call-for-a-distributed-web/

Kamkar, Samy. 2010. Evercookie. *Samy.Pl.* September 20, 2010. https://samy.pl/evercookie/. Archived version available in Internet Archive: https://web.archive.org/web/20170930221445/https://samy.pl/evercookie/

Klein, Martin, Herbert Van de Sompel, Robert Sanderson, Harihar Shankar, Lyudmila Balakireva, Ke Zhou, and Richard Tobin. 2013. Scholarly Context Not Found: One in Five Articles Suffers From Reference Rot. *PLoS ONE* 9 (12): e115253.

Koehler, Wallace. 2004. A Longitudinal Study of Web Pages Continued: A Consideration of Document Persistence. *Information Research* 9 (2). http://www.informationr.net/ir/9-2/paper174.html.

Krishnamurthy, Balachander, and Craig E Wills. 2009. Privacy Diffusion on the Web. A Longitudinal Perspective. In *WWW 2009, Proceedings of the 18th*

International Conference on World Wide Web, Madrid, April 20–24, 2009. http://citeseerx.ist.psu.edu/viewdoc/download?doi=10.1.1.232.3038&rep=-rep1&type=pdf

Lawrence, Steve, David M. Pennock, Gary William Flake, Robert Krovetz, Frans M. Coetzee, Erik Glover, Finn Årup Nielsen, Andries Kruger, and C. Lee Giles. 2001. Persistence of Web References in Scientific Research. *IEEE Computer* 34 (2): 26–31.

Liebler, Raizel, and Liebert June. 2013. Something Rotten in the State of Legal Citation. The Life Span of a United States Supreme Court Citation Containing an Internet Link (1996–2010). *Yale Journal of Law and Technology* 15 (2): 1–39.

Lyman, Peter. 2002. Archiving the World Wide Web. Council on Library and Information Resources. http://www.clir.org/pubs/reports/pub106/web.html. Accessed September 29, 2017. Archived version available in Internet Archive: https://web.archive.org/web/20140706052301/https://www.clir.org/pubs/reports/pub106/web.html

Maclay, Kathleen. 2009. Web Privacy Report Finds Widespread Data Sharing, "Web Bugs". *UC Berkeley News*, July 2, 2009. http://www.berkeley.edu/news/media/releases/2009/06/02_webprivacy.shtml. Archived version available in Internet Archive: https://web.archive.org/web/20160624022704/http://www.berkeley.edu/news/media/releases/2009/06/02_webprivacy.shtml

Masanes, Julien. 2005. Web Archiving Methods and Approaches: A Comparative Study. *Library Trends* 54 (1): 72–90.

Massicotte, Mia, and Kathleen Botter. 2017. Reference Rot in the Repository: A Case Study of Electronic Theses and Dissertations (ETDs) in an Academic Library. *Information Technology and Libraries* 36 (1): 11–28.

Mayer, Jonathan R., and John C. Mitchell. 2012. Third-Party Web Tracking: Policy and Technology. In *IEEE Symposium on Security and Privacy*, 413–427. San Francisco, May 20–23, 2012.

McCullough, Brian. 2014. On the 20th Anniversary, an Oral History of the Web's First Banner Ads. Accessed September 28, 2017. http://www.internethistory-podcast.com/2014/10/the-webs-first-banner-ads/. Archived version available in Internet Archive: https://web.archive.org/web/20170706102419/; http://www.internethistorypodcast.com/2014/10/the-webs-first-banner-ads/

McDonald, Aleecia M., and Lorrie Faith Cranor. 2011. A Survey of the Use of Adobe Flash Local Shared Objects to Respawn HTTP Cookies. *Cylab.Cmu.Edu*. CyLab, Carnegie Mellon University, January 31, 2011.

Merriam-Webster. n.d.-a Record. Accessed September 23, 2017. https://www.merriam-webster.com/dictionary/record

―――. n.d.-b Register. Accessed September 23, 2017. https://www.merriam-webster.com/dictionary/register

Montulli, Lui. 2013. The Irregular Musings of Lou Montulli: the Reasoning Behind Web Cookies. *Web.Archive.org*, May 14, 2013. Archived version available in Internet Archive: https://web.archive.org/web/20130627180619/http://www.montulli-blog.com/2013/05/the-reasoning-behind-web-cookies.html

Mowery, Keaton, and Hovav Shacham. 2012. Pixel Perfect: Fingerprinting Canvas in HTML5. Proceedings from the Web 20 Workshop on Security and Privacy.

Nielsen, Janne. 2016. *Using Web Archives in Research: An Introduction*. 1st ed. Aarhus: NetLab.

Nikiforakis, Nick, Alexandros Kapravelos, Wouter Joosen, Christopher Kruegel, Frank Piessens, and Giovanni Vigna. 2013. Cookieless Monster: Exploring the Ecosystem of Web-Based Device Fingerprinting. In *Proceedings from the 2013 IEEE Symposium on Security and Privacy*, 541–555. Washington, DC, May 19–22, 2013.

Nyvang, Caroline, Thomas Hvid Kromann, and Eld Zierau. 2017. Capturing the Web at Large: A Critique of Current Web Referencing Practices. In *Proceedings from the Researchers, Practitioners and their Use of the Archived Web Conference* (RESAW2 London, 2017). https://doi.org/10.14296/resaw.0002

Right to Be Forgotten. n.d. *The Guardian*, 2015–2017. https://www.theguardian.com/technology/right-to-be-forgotten. Archived version available in Internet Archive: https://web.archive.org/web/20171001191902/https://www.theguardian.com/technology/right-to-be-forgotten

Roesner, Franziska, Tadayoshi Kohno, and David Wetherall. 2012. Detecting and Defending Against Third-Party Tracking on the Web. In *Proceedings of the 9th USENIX Symposium on Networked Systems Design and Implementation*, San José, CA, April 25–27, 2012.

Schneider, Steven M., and Kirsten A. Foot. 2004. The Web as an Object of Study. *New Media & Society* 6 (1): 114–122.

Schostag, Sabine, and Eva Fønss-Jørgensen. 2012. Webarchiving: Legal Deposit of Internet in Denmark, a Curatorial Perspective. *Microform & Digitization Review* 41 (3–4): 110–120.

Shen, Catherine. 2014. Tracking the Trackers: Investigators Reveal Pervasive Profiling of Web Users. *Princeton.Edu*, November 5, 2014. https://www.princeton.edu/news/2014/11/05/tracking-trackers-investigators-reveal-pervasive-profiling-web-users. Archived version available in Internet Archive: https://web.archive.org/web/20170916123406/https://www.princeton.edu/news/2014/11/05/tracking-trackers-investigators-reveal-pervasive-profiling-web-users

Simpkins, Lindsay, Xiaohong Yuan, Jwalit Modi, Justin Zhan, and Li Yang. 2015. A Course Module on Web Tracking and Privacy. In *Proceedings of the 2015 Information Security Curriculum Development Conference*, Kennesaw, GA, October 10, 2015

Soltani, Ashkan, Shannon Canty, Quentin Mayo, Lauren Thomas, and Chris Jay Hoofnagle. 2009. Flash Cookies and Privacy. *SSRN*, August 10, 2009. https://papers.ssrn.com/sol3/papers.cfm?abstract_id=1446862

Stringer, Nick. 2012 All About… The EU Cookie Law. *Campaign*, April 19, 2012. https://www.campaignlive.co.uk/article/eu-cookie-law/1127629

Tene, Omer, and Jules Polonetsky. 2011. To Track or "Do Not Track": Advancing Transparency and Individual Control in Online Behavioral Advertising. *SSRN*, September 1, 2011, https://papers.ssrn.com/sol3/papers.cfm?abstract_id=1920505##

Thomas, Arthur, Eric T. Meyer, Meghan Dougherty, Charles Van den Heuvel, Christine McCarthy Madsen, and Sally Wyatt. 2010. Researcher Engagement with Web Archives: Challenges and Opportunities for Investment. Jisc. Joint Information Systems Committee.

Tracker Tracker. 2017. Digital Methods Initiative Amsterdam. Accessed September 19, 2017. https://wiki.digitalmethods.net/Dmi/ToolTrackerTracker. Archived version available in Internet Archive: https://web.archive.org/web/20160304015338/https://wiki.digitalmethods.net/Dmi/ToolTrackerTracker

Turow, Joseph. 2012. *The Daily You. How the New Advertising Industry is Defining Your Identity and Your Worth*. New Haven: Yale University Press.

Webster, Peter. 2013. Political Party Web Archives—UK Web Archive Blog. *Blogs.Bl.Uk*, December 11, 2013. http://blogs.bl.uk/webarchive/2013/12/political-party-web-archives.html. Archived version can be reconstructed by using Memento Time Travel: http://timetravel.mementoweb.org/reconstruct/20171024154310/http://blogs.bl.uk/webarchive/2013/12/political-party-web-archives.html

Weinberger, Amy. 2011. The Impact of Cookie Deletion on Site-Server and Ad-Server Metrics in Australia. *comScore*, February 3, 2011. http://www.comscore.com/Insights/Presentations-and-Whitepapers/2011/The-Impact-of-Cookie-Deletion-on-Site-Server-and-Ad-Server-Metrics-in-Australia-An-Empirical-comScore-Study?&cs_edgescape_cc=DK

Zittrain, Jonathan, Kendra Albert, and Lessig Lawrence. 2014. Perma: Scoping and Addressing the Problem of Link and Reference Rot in Legal Citations. *Harvard Law Review Forum*, March 17, 2014. https://harvardlawreview. org/2014/03/perma-scoping-and-addressing-the-problem-of-link-and-reference-rot-in-legal-citations/. Archived version available in Internet Archive: https://web.archive.org/web/20170914162229/https://harvardlawreview. org/2014/03/perma-scoping-and-addressing-the-problem-of-link-and-reference-rot-in-legal-citations/

You Press the Button, We Do the Rest. Personal Archiving in Capture Culture

Jacek Smolicki

1 Introduction

Given the contemporary abundance of personal technologies, we engage in capturing ever more detailed aspects of everyday lives while simultaneously exposing them to involuntary capture performed by third parties. In the context of ubiquitous technologies we have become at once *micro-archivists* and *micro-archives*, or, differently put, *subjects* and *objects* of pervasively radiating micro-temporal mechanisms of capture, mediation and storage of our digitized life-bits. In this chapter, I intend to problematize the recent transformations in the domain of technologies of record in relation to Bernard Stiegler's concepts of *mnemotechniques* and *mnemotechnologies*. The transition from mnemotechniques to mnemotechnologies signifies a move from low-tech, controllable and manual techniques of handling one's memory, its accumulation and processing (a craft, or *ars memoriae*, if you will) to large-scale, automated mnemonic industries

J. Smolicki (✉)
Malmö University, Malmö, Sweden
e-mail: jacek@smolicki.com

© The Author(s) 2018
A. Romele, E. Terrone (eds.), *Towards a Philosophy of Digital Media*,
https://doi.org/10.1007/978-3-319-75759-9_5

(*industria memoriae*) which, while providing customers with immediate and easy solutions for capturing and organizing their accelerating flows of memories, simultaneously benefit from them economically. Today we can observe a vast economic and aesthetic reconfiguration of what it means to be recording and preserving memory, or rather capturing, which, as I argue, in the domain of network technologies has been lately eroding the use and significance of the term "recording". I demonstrate this transition by looking at the way that *snap-shooting* (the early practice of amateur photography enabled by the first mass-produced cameras such as the Brownie) has mutated in the context of ubiquitous computing and mnemotechnologies such as Facebook. What follows this discussion on the confluence of voluntary and involuntary capture and archiving is a brief overview of counter-practices motivated to impede this current techno-political condition. Drawing on the concept of the post-digital, which to some extent correlates with the notion of mnemotechnique through its call for a return to manual forms of engagements with technology, as an epilogue to this chapter I point towards an alternative itinerary for how capturing and archiving practices can be rethought today.

2   From Recording to Capturing

The verb "to record" originated from the Latin *recordi*. It consist of two components: *re-*, to restore, and *cordari*, a heart.[1] In ancient times a common belief was that it is the human heart (not mind), where memory and thus mnemonic capacities are actualized and taken care of. In his writing, Aristotle often referred to the heart as the base of wisdom, situated reasoning and practical thinking. In contrast, the brain was to serve as a "cooling" device, an instrument that controls and rationalizes the processes that the heart undertook (Gross 1995). We do not have to look too far to find remnants of this belief. Consider the phrase "learning by heart", which signifies such a process of internalizing knowledge. Thus, according to this etymological perspective, recording stands for active and thoughtful processes of internalizing, ordering thoughts, remembering, calling them to mind, rethinking and being attentive. In other words, recording signifies a deliberate construction and retention of memory, a process in which one's mental and corporeal agencies operate in tandem.

I would like to suggest that today the term recording and its meaning have become challenged by another term: "capturing". The term is used increasingly, not only as an equivalent of taking visual snapshots, but also in relation to other types, formats and categories of personal data. A number of technological devices, gadgets and services dedicated to the storage, organization and sharing of personal memory draw on the term in their advertising strategies. For example, Evernote encourages you to "capture what's on your mind," the popular life-logging wearable camera and mobile app Narrative Clip invites you to "capture authentic photos and videos effortlessly." Some other corporations and start-ups incorporate the term into the very name of their service, such as Capture-app, a photo storage service automating personal photography and video collections, or Kapture, an audio recording wristband enabling the effortless capture of audio snippets from daily life. As it is seen quite clearly in some of these cases, the verb capture is often followed by an indication of its effortless, automated character, simple use and general user-friendliness which all make the term mean something much different from recording.

The terms "capture" and "capturing" are etymologically related to the Latin *capio* and *captum*. They mean precisely "to take" or "to seize." The Online Etymology Dictionary[2] links capture to Latin *captura* and *captus*, which translate into "taking captive": an act of taking control of something. Thus, while on the one hand capturing connotes a voluntary act of exercising one's individual agency by taking control of something (in this case a process of visually documenting everyday life), it also signifies an imposition of some force, and consequently, an act of taking something or someone captive.

A simple glance at the cluster of visual images resulting from a Google search points at this twofold meaning of the term "capture." On the one hand, we are presented with images of hand-held cameras and people pressing the shutter in what appears to be a voluntary act of taking a picture. Other pictures show us a human (or a mouse) confined in a box or surrounded by an array of hostile traps, simultaneously luring the potential victim and blocking any other navigation than the one possible within the confines. One could say that, in a somewhat banal way, these contrasting images, crudely juxtaposed by Google algorithms, mirror the co-constituent dynamics of what I have been calling capture culture; dynamics resulting from the intersecting planes of ever more easy, (over-)simplified and seemingly empowering modes of self-surveilling and accounting for everyday life, but also an inescapable exposure to strategically automated, "forensic" media technologies which imperceptibly perform involuntary capture and a subsequent archiving of personal data.

One could ask here how capturing is related to archiving, a key notion to this article. In taking a picture via one of the many popular smartphone applications (such as Instagram), we may primarily intend to capture a moment without considering this act in terms of archiving. Nevertheless, the captured image often becomes part of a larger collection, and since it is labeled and tagged with additional metadata (entered intentionally or inserted automatically by the operating system, software or an application used for capturing), it becomes equipped with some degree of indexical information. It can be said that this insertion of indexical information turns the captured into the archived. Wolfgang Ernst (2012) goes even further, arguing that all digital content (and thus a digitally captured image, video, sound file) is in itself a kind of (techno)archive. Since a single digital file consists of a set of alphanumeric characters, every such character, frame or even pixel can be discretely addressed and recalled, just like a properly indexed document stored on the shelves in a "traditional" archive or personal filing folder. Besides this, there are other arguments that enable us to speak of a close relationship between digital capturing and archiving. While the results of our intentional acts to capture our everyday lives extend into digital archives compiled beyond our perception and intention, devices that are not originally designed to capture and archive our daily life activities increasingly do so. Consider smart objects and the Internet of Things, where objects such as TV sets, devised to serve people by performing anticipated functions, may suddenly turn into capturing and archiving agents simply through their being connected to the internet.[3] The ubiquitous capture and archiving of people's everyday lives taking place alongside a multitude of vectors and directions is the overall cost of connectivity, immediacy and automation, which in turn can be seen as some of the key attributes characterizing the current phase of network computing development, or "mnemotechnnologies," to use Bernard Stiegler's term.

3 Mnemotechniques and Mnemotechnologies

Stiegler's archaeology of reflexivity, as he himself defines his scholarship, revolves closely around relationships between humans, memory and technics (1998). Overall, Stiegler's analysis of these relationships builds on an observation that "the human evolves through externalizing itself in tools, artefacts, language and technical memory banks" (Stiegler 2010, 65). Thus, technology cannot be seen as some autonomous entity simply added to a non-technological human existence. On the contrary, the technical is always an essential dimension and condition of this existence. Consequently agreeing with this inseparable correlation between people and technologies, human and technological evolutions have to be seen as processes that are always intertwined. Stiegler calls this condition "originary technicity" or alternatively "originary prostheticity" (Stiegler 1998, 153).

Writing about recent technological transformations that affect this relationship, Stiegler proposes a differentiation into two particular modes of living and becoming with techne: *mnemotechniques* and *mnemotechnologies* (2010). These two notions describe different kinds of complexity characterizing relationships between human mnemonic capacities (*mneme*) and technological prostheses (*hypomnemata*) through which humans exteriorize memory and knowledge.

What are mnemotechniques and how do they differ from mnemotechnologies? Based on my reading of Stiegler in relation to capturing and archiving practices, mnemotechniques can be seen as intentional acts of externalizing memory and knowledge. Intentionality, connected to the notions of agency and freedom (Verbeek 2011, 13), signifies here the will and intention to act in a specific way. The processual dynamics of mnemotechnical practices are comprehensible and controllable by individuals (or communities) directly; in other words, people voluntarily engage in these practices. Mnemotechniques evolve in line with specific motivations which guide the creation and organization of the resulting outcome, such as a mnemonic trace. Mnemotechniques are a form of craft, which, by Tim Ingold's definition, we might see as a manual practice comprising "observational engagement and perceptual acuity that allow the practitioner to follow what is going on, and in turn to respond to it" (2013, 4). Mnemotechnique involves a binding connection between one's memory and its externalized form. It results in an associated mnemonic milieu where the bond between human subjects (their consciousness), the technologies through which they experience and register the world, and the material records of their experiences tend to hold strong over time. A simple example of a mnemotechnique on a personal level could be note-taking performed by hand, or a drawing that results in an artifact, a material trace marked by a distinctive style, simultaneously a signifier of a close correspondence between the technological and the human. Although the primary intention of mnemotechniques is to record and transmit the memory of a given experience, every activity that results in the externalization of knowledge into a (technical) artifact can potentially become a vector of memory, and hence, a non-intentional type of mnemotechnique (for example, a production of a wheel, flint, or pottery is both a materialization of a person's memory and knowledge and an encapsulation of this knowledge and experience into a physical trace or artifact).

In contrast to mnemotechniques, mnemotechnologies (which to indicate their scale Stiegler also denotes in terms of "global mnemotechnics" ([2003]) are technologies that take control of and automate people's memory practices, implicating them into other relations beyond the integral connection with the subject or community. As examples of such mnemotechnologies, Stiegler refers in particular to mass media, radio, television and mainstream cinematography; in other words, media infrastructures historically identified as strictly divided into phases of production and consumption, and, subsequently, a sharp distinction of media practitioners into producers and consumers. In his opinion, these mass media and technologies provide receivers with a series of fabricated and pre-established, mass-scale audio-visual experiences and narratives, effectively paralyzing all opportunity to construct a meaningful relationship with the world. Stiegler describes the impact of mnemotechnologies in terms of the *proletarianization* of individual and collective mental power, attention, and sensibility (2016). Much as how industrial capitalism proletarianized labor and blocked people's capacity to draw meaning and satisfaction from what they make, consumerist capitalism proletarianized people's ability to care about and pay attention to what is meaningful in their lives. In other words, just as the behavior of the worker at the dawn of the nineteenth century was contingent on the service of an industrial machine, the behavior of the contemporary consumer has been "standardised through the formatting and artificial manufacturing of his desires" (2011, no page).

"Adoption" and "adaptation," terms that recur quite often in Stiegler's work, may be seen as helpful in further clarifying the difference between mnemotechniques and mnemotechnologies. Mnemotechnologies make people *adapt*, somewhat passively, to their black-boxed, internal rules, automated programming, as opposed to making themselves *adoptable* to the needs, priorities, temporalities and rhythms of the everyday lives they enter. This intractable and intrusive nature of mnemotechnologies suggests an obvious comparison to Adorno's and Horkheimer's concept of the culture industry (1972). As a matter of fact, Stiegler quite explicitly builds on this concept (2011). However, in contrast to these authors, Stiegler does not stop at an entirely pessimistic vision. What can be seen to motivate his diagnosis of contemporaneity in terms of the mnemotechnological production of cultural "malaise" (2010), "short-termism," and "symbolic misery" (2014) is in fact a conviction that this situation can (and must) be critically reconfigured. This "re-enchantment of the world," as he calls it, points toward acts of rehabilitating and reclaiming mnemotechnical qualities in our co-existence with technologies, not in the realm of high art, but in how we interact with technologies on a daily basis. As Stiegler maintains, this renewed approach can be achieved today through the pervasiveness of new media, digital and network technologies such as the web and personal communication devices. Somewhat echoing the early techno-enthusiasm surrounding the emergence of the web (and particularly web 2.0), Stiegler perceives democratization of power and opening of participation in media production as crucial forces to restore a chance to liberate "hypomnesic memory from its industrial function" (2010, 85). Such a depiction of the contemporary media landscape is obviously problematic, and rather dated, considering the range of critiques of digital and network media technologies published in recent years (for instance, van Dijck 2007, 2013), of which some even borrowed directly from Stiegler's discussions (see Lovink 2011).

I suggest that the concept of mnemotechnology can in fact be extended further into contemporary mainstream media technological devices, systems and services that exteriorize, aggregate and circulate personal memories. Thus, I use the term "mnemotechnology" as a critical one, extrapolating it also onto mainstream network technologies and devices, smartphones, data-aggregation and social-media platforms, some of which I will discuss in the following section of this chapter. The following words from Stiegler's (2010, 67) writing on mnemotechnological systems appear particularly applicable in this context:

> Originally objectified and exteriorized, memory constantly expands technically as it extends the knowledge of mankind; its power simultaneously escapes our grasp and surpasses us, calling into question our psychical as well as our social organization. This is particularly apparent in the transition from mnemotechniques to mnemotechnologies from individual exteriorizations of memory functions to large-scale technological systems or networks that organize memories.

As we learn from this excerpt, mnemotechnologies dissociate the mediated memory (and know-how skills) from an individual. One's memory and knowledge become disassembled and subsequently reassembled in accordance with externally developed mechanisms of "distantly operating service industries" that "network people's personal memories [. . .] control them, formalize them, model them, and perhaps destroy them" (Stiegler 2010, 68). Consequently, and in contrast to mnemotechniques, mnemotechnologies lead to *dissociated* mnemonic milieux in which memories become increasingly displaced and disjointed, forming imperceptible realms, "black-boxed" enclosures affected by terms and conditions often non-amendable by the subject on whose memory they scavenge. Mnemotechnologies release their users from the effort of actively recording, recalling and putting knowledge into action, effectively turning them into the operators of ready-made systems that input and output, process and deliver knowledge and memory in an increasingly passive way (Lovink 2011, 25).

4   From the Mass of Comparative Lumber to the Techno-Value of Data Excess

How are these mnemotechnological dynamics realized in the context of today's media technological landscape more specifically? The title of this chapter has been inspired by the slogan advertising the first, portable consumer-dedicated cameras launched by Kodak at the dusk of the nineteenth century. The camera came with a roll of film which, after completion, the photographer was invited to send back to the company for the subsequent processing and development into physical photographs. Based on the same principle, a couple of decades later, Kodak released the Brownie, its first truly mass-produced photo-camera for a wider public. Within the first year of its appearance on the market, 100,000 of these devices were sold. It is also with the emergence of this camera that the term "snapshot" (previously used solely in relation to firing guns and hunting) was first introduced to describe a photographic style and aesthetics involving a spontaneous and quick push of a button in response to the moment. This new aesthetic practice of snap-shooting released photography from its dependence on the long and complex professional procedures, stationary, bulky, and expensive equipment, facilities, and locations it typically relied on in professional portrait studios. While on the one hand the democratization of the photographic medium was seen as an entirely positive phenomenon doing away with the monopoly over visual representation enjoyed by privileged individuals, artists and historians, on the other hand it invoked concerns and even fear, precisely among historians and archivists. The undisciplined production of images made them speak of what today we could describe as big data, data excess or exhaust. One hundred years ago, in one of the earliest books that attempted to tackle this problem, historians H.D. Gower, Louis Stanley Jast and William Whiteman Topley wrote:

> One of the outstanding characteristics of our time—as contrasted with the past—is its tendency towards the elimination of waste of power—the economical direction of human activities so as to increase their productivity and effectiveness—the utilization of what have heretofore been regarded as "waste products." To the engineer it is abhorrent that any energy be allowed to run to waste.

But in the domain of photography the amount of horse power running to waste is appalling—and all for lack of a little system and co-ordination. Shall this be allowed to continue? Shall the product of countless cameras be in the future, as in the past (and in large measure today) a mass of comparative lumber, losing its interest even for its owners, and of no public usefulness whatever? This is a question of urgency. Every year of inaction means an increase of this wastage. (Gower et al. 1916, 6)

While the quote above signifies a certain care for the historical legacy the first generation of amateur photographers was going to leave behind, it also reveals a strong inspiration of Taylorism, at the time an increasingly popular theory of management applied within industrial production. Briefly put, Taylorism (named after its inventor, mechanical engineer Frederick Winslow Taylor) aimed to standardize best practices across sites of industrial production and manufacturing to improve the overall work flow, management and economic efficiency. "The Camera as Historian," which was the title of the handbook proposed by the said historians, suggests that the tendrils of the belief in scientific management at the time also extended to the realms of archiving, cultural, aesthetic production, heritage, and historical imagination. A systemic and rigorous application of rules according to which people were encouraged to engage in a photographic practice was proposed by them as a possible solution to the growing "mass of comparative lumber." Effectively, their joint text functioned as a handbook for how to set up and engage in disciplined photographic surveys that could conduce to historically relevant depictions of everyday life from within its currents, and thus yield a meaningful and informative legacy for posterity. Taking into consideration the present ubiquity of capturing devices, how are these early debates informative to the current condition and amateur snap-shooting practices?

The premonition encapsulated in the cited excerpt can be said to have been fulfilled precisely 100 years later. Current statistics reveal that an average upload of photographs on Facebook reaches about 350 million, daily.[4] Some 95 million are posted daily on Instagram.[5] These numbers are still rather insignificant compared to the amount of "comparative lumber" circulated via Snapchat, reaching up to 3 billion daily.[6] Besides the ease of taking a snapshot in the age of capture culture, these numbers also account for another unprecedented phenomenon of the present moment: an image has today transformed from a signifier of the past, a trace of a remarkable moment intended to endure the passage of time, to a basic unit for communicating the most mundane, instantaneous and fleeting, and not necessarily beyond this very moment. The longevity of a photographic image, almost irremovable today from the context of network media technologies, has radically shrunk. However, what on the surface appears to constitute a continuous "flow through the present" (Bogost 2010, 28) at its depth imperceptibly accumulates into carefully anticipated archival strata. Contemporary amateur photography practices (a remediation of earlier snap-shooting activities) are today, as Gower and his peers would wish, significantly influenced and determined to feed archives. However, not ones characterized by long-term, cultural and historical significance, but rather short-term, economical and market value.

5   Ubiquitous Capture and Involuntary Sedimentation

Sarah Kember calls this condition "ubiquitous photography" (2014). She suggests that the recent expansion of network technologies and online social-media services radically modified vernacular photographic practices. It is ever harder to speak of them in terms of ideas like bottom-up culture, user-centrism, rethought historical imagination and autonomy from institutional frameworks, aesthetics and ideologies. Transformed radically by ubiquitous computing, everyday practices of photographic documentation have become an intrinsic layer of a larger apparatus of "ambient intelligence" constituted by a network of technological infrastructures, hardware and software monitoring, aggregating and processing personal data (not only images) mediated via a growing array of personal devices such as smartphones, wearable lifelogging cameras, self-tracking gadgets to name a few. While Kember describes amateur photographic practices taking place across mainstream mnemotechnologies in terms of "contested ground," web-security expert Bruce Schneier goes even further, calling the whole territory of the web 2.0 a "feudal land" (Schneier 2015). He sees big corporations that presently orchestrate the market of network technologies (for example Apple, Google, Facebook, Amazon) as "analogous to the feudal lords" where "we are their vassals, peasants, and—on a bad day— serfs. We are tenant farmers for these companies, working on their land by producing data that they in turn sell for profit" (Schneier 2015, 58). This is why today, in the context of big-data economy, there is "no category or amount of data that is ruled out a priori from the perspective of those who seek to mine it for unknowable and unpredictable patterns" (Andrejevic 2013, 96). The overabundance of images invoking fear among historians 100 years today, to some, invokes a prospect of profit. Thus, websites and social-media services are being deliberately engineered and designed to influence their users to account for their everyday life as thoroughly as possible, so as to increase the chance for potential profit from the outcomes of the algorithmic operations performed upon them, and to provide enough data points for a precise targeting of the subject, and thus to perpetually stimulate data production and consumption. To paraphrase Vilém Flusser, mnemotech-nologies concerned with personal data aggregation (such as Facebook) turn the activities of their users (or serfs, as Schneier would prefer) into their very "function" (Flusser 2011, 10).

The context of network technologies and social media services, ever more intrusively working their way into practices of documenting every-day lives, make these practices and their products an inevitable part of a contract with their providers. Despite being marketed as free of charge, a number of mainstream mnemotechnologies involved in capturing every-day life experiences have a strong transactional component to them. In order to easily, efficiently capture and mediate memories of everyday life and to subsequently gain access to their display (and analysis), the user accepts the possibility of disclosing this data to an unknown gaze and procedures. In photography (as with any other kind of data) the subject of this transaction is not necessarily the content, but also, or often pri-marily, its surroundings, which is to say, its meta-data (location, time, information on people viewing the content, liking it, or being tagged in it). The transaction often happens without full consent or with only lim-ited understanding of the potential implications and purposes that might be made of data points surrounding, in this case, the image and people's interactions with it (captioning, commenting, liking, distributing). The principles and complexities of such mnemotechnological transactions are often obscured in lengthy terms of agreement that, as some studies show, only about 7% of users bother to take a look into.[7]

As might be anticipated, Facebook lends itself as perhaps the most noteworthy example of a platform persistently working towards an increase of financial benefits from the myriad of communication streams that it facilitates. The history of Facebook might be seen as a gradual move from it being an enabler of smooth communication to a quasi-archival apparatus scavenging on the users' capturing, or what in the past we would have called mnemonic practices. In December 2011, Facebook changed its layout from a less structured database of updates and notifications to the timeline, a chronologically evolving collection of images and textual information about a person's everyday life. While older versions of Facebook were mostly concerned with news feeds, quick updates, and notifications from friends, the timeline added the feel of compiling a personal archive, or a chronicle of one's life (van Dijck 2013). The introduction of the timeline also marks a moment when memory, emotion, and nostalgia become part of the Facebook experience and an integral component of their business strategy, expanding its function way beyond a mere communication service and news feed. As van Dijck points out, the timeline structure encouraged users to post pictures from "the pre-Facebook days of their youth—a baby picture, family snapshots, school classes, old friends, college years, wedding pictures, honeymoon—and thus experience content in terms of their life's story" (van Dijck 2013, 55). While revamping the service's architecture, Facebook also started updating its privacy policy. From then on, all content uploaded on the timeline was, by default, set as public with privacy as an option. Only a month after the timeline was introduced, Facebook began extensive collaboration with external companies, whose advertisements started cluttering users' profiles. Thus, the introduction of the timeline can be seen as pivotal for Facebook on its way to unprecedented commercial gains and dominance on the data market. While the timeline offered the user a chance to neatly organize his/her flow of data and decide on its openness and disclosure, it simultaneously established a more efficient infrastructure for covert data aggregation and storage. As van Dijck highlights, the uniformed and streamlined design ultimately benefited advertisers and other third parties. They could now more efficiently extract content from the users' daily interactions, which became far more commensurate and comparable than they were before the timeline was introduced. In other words, while the front end of Facebook captivated its users with its visually attractive and relatively simple and user-friendly structure, enabling them to construct what seemed to them an associated mnemonic milieu, an imperceptible, dissociated mnemonic milieu was simultaneously being accumulated on the deep end of the timeline.

With the timeline structure, Facebook also started offering its users the chance to copy the profile by downloading a zipped file that would seemingly include all the archived information in one's Facebook profile. As Austrian law student Max Schrems has proved, however, this zipped file is only a fraction of what Facebook really archives. After submitting a request under the European "right to access" law, he made Facebook disclose all the information that the corporation had collected on him. Facebook sent him a CD with a 1,200-page PDF document containing far more substantial data than the generic archive offered for download (which, according to Schrems, amounts to only about 29% of the aggregated data). The data that was not included in the Facebook service concerned, for instance, the "like" button function, tracking on other webpages, results of face recognition, videos, posts launched on other users' timelines, pages viewed while signed into Facebook, and previously deleted content, such as tags or comments.[8]

An in-depth study of this hidden archival sedimentation dissociated from the user's perception has recently been pursued by Share Lab, a Serbian data investigation lab exploring various technical aspects of the intersections between technology and society.[9] According to their study, personal data that Facebook is interested in recapturing and archiving for its purposes comes from four major sources: actions and behaviors, profile and account, digital footprint, and outside of the Facebook domain. Actions and behaviors are primarily activities performed voluntarily, such as uploading images, or visiting pages by clicking on links embedded or posted by others. While the number of these actions seems quite low, the effect of each is captured and subsequently discretized into between four and six other data sets, which altogether amass up to about 80 different categories of information that Facebook's algorithms are thereafter able to harvest, archive, analyze and sell. The second group, profile and account information, mainly consists of voluntarily typed information, usually inserted by the user and which remains relatively static compared to the flux of mediated memories that are the actions and behaviors regularly published and shared on the timeline. The third source of personal data is the digital footprint, which is information gathered from mobile phones, laptops, and desktop computers from which users access Facebook and, more generally, the web. This is an example of how Facebook can easily move beyond its framework to harvest personal data. This mode of personal data archiving is involuntary compared to the intentionally logged data discussed above. Thus, it is alternatively called a "passive digital footprint." It is enabled by trackers such as cookies; cookies being small pieces of software sent from a website and stored on the user's computer or mobile device to capture their user's various actions. Trackers gather and send back information such as the identity of a user's computer and his/her movements across the web. One need not even be directly active on Facebook to become the subject of such tracking and archiving procedures. Instead of us visiting the Facebook page, through the pervasive surveying or "surveilling" agents, as we could alternatively describe trackers, Facebook can "visit" us whenever we browse through a website that has such trackers incorporated into its architecture, including advertisements and "like" buttons which, as of March 2015, populated about 13 million pages worldwide.[10] Facebook trackers collect and store information on its users whether or not they are logged in. This points at the fourth category: data collected from outside Facebook. Since Facebook owns a range of other online services (for example WhatsApp, Instagram, Atlas, to name a few), or is in close partnership with other companies (such as Acxiom, one of the largest players in US database marketing), the data that these services harvest is subsequently channeled into Facebook databases or exchanged for data gathered directly by Facebook.[11] Moreover, several years ago, Facebook introduced "data cookies," which are installed on the individual's computer browser and remain there for two years after clicking the like button. This allows Facebook to continuously collect data from the browser, expanding its tracking dominion over those who do not use Facebook or who have decided to deliberately opt out.

6   Modes of Counteracting the Involuntary Archive

Questions arise as to what kind of moves can be initiated in order to impede or challenge this dominion of mainstream mnemotechnological industries. Some scholars suggest that there is no life outside the effects of mainstream network technologies (Lovink 2011; Deuze 2012). Whether we want it or not, our daily choices, movements, decisions and memories are subject to capture and potential archiving, with the procedures by which this is achieved being deliberately kept concealed from our immediate attention. In 2009 Gordon Bell, the father of so called life-logging, "a form of pervasive computing, consisting of a unified digital record of the totality of an individual's experiences, captured in a multimodal way through digital sensors and stored permanently as a personal multimedia archive" (Dodge and Kitchin 2007, 2) argued that we are at a cusp of an age where "it will require a conscious decision (or a legal requirement) *not* to record [emphasis added] a certain kind of information in a certain time or place—the exact reverse of how things are now" (Bell and Gemmell 2009, 8). As a response to that state he proposed a radical form of embracing digital, fully automated technologies to capture the entirety of one's life, and by doing so, (re)gaining full control over the way one's daily life account is constructed. In what can be seen as an ultra-enthusiastic response to the so called "digital revolution," Bell believed that human agency should be entirely written out of the process, and replaced by passively operating sensors and monitoring devices. In his vision, the proverbial button from the title of this chapter would no longer be needed. The act of capturing one's everyday life would be taking place passively, continuously, latently and imperceptibly. One could say that to some degree his vision has been fulfilled in practices of Quantified Self (QS). Described by Dawn Nafus and Jamie Sherman (2014, 1784–94) as a form of "soft resistance" to big-data mechanisms, QS members use passive sensing and tracking technologies to monitor and capture various aspects of their daily routines. Narrative Clip, a wearable camera passively documenting the daily life of its owner can be seen as another instance of an attempt to take control over one's flow of mediated memories through an appliance of a passive, fully automated technical device.[12]

While some advocate for a full submission to the capturing possibilities brought about with the proliferation of cheap, consumer-dedicated technologies, others move in an entirely opposite direction. Here I have in mind a range of counter-surveillance practices emerging lately, primarily among creative media practitioners, activists and artists. The practice of *obfuscation* can be seen as a specific, tactical way of counteracting the working of mainstream mnemotechnological services by introducing a significant amount of misinformation, noise, ambiguity and confusion into one's digital data practices. In their latest book, scholars, artists, and programmers Finn Brunton and Helen Nissenbaum (2015) describe the goal of obfuscation practices as "mitigating and defeating present-day digital surveillance." The goal of obfuscation is to disable, or at least, temporarily deny involuntary observation and capture of personal data. It is performed through creating many plausible, ambiguous, and misleading signals within which the information we want to conceal can be lost. Assuming that every signal sent and received via network technologies can be spotted, tracked, captured and cataloged, proponents of obfuscation suggest that instead of limiting, or withdrawing from the use of mnemotechnologies (which might be impossible) one should deliberately add to existing signals. Generating a bundle of random signals amasses a dense and noisy plethora of misleading patterns, among which one's real presence may be temporarily concealed and unidentifiable. Obfuscation is not about the mindful reduction of one's involvement in consuming and circulating media content; on the contrary, it is about even more intensely engaging in digital data flow and generating even more "comparative lumber" than mnemotechnological industries would ever expect from us. Similarly, it is not about openly confronting and counteracting mechanisms of involuntary capture and archiving, in a literal sense, but tactically utilizing, playing with and subverting the fact of their inevitable presence in our everyday lives (Brunton and Nissenbaum 2015, 55).

7   Conclusion: Towards Post-Digital Mnemotechniques

In the first section of this chapter I suggested that we are witnessing a certain shift in the way we attend to recording memories and documenting our everyday lives. This shift was expressed through a juxtaposition of the term "recording" with the term "capturing", the latter, as I argued, being ever more expansively used today in relation to the ubiquity of digital and network technologies. The shift from recording to capturing can be articulated alternatively in terms of a move from the qualitative to quantitative, active to passive, manual to automated modes of constructing and mediating personal memories. In a similar way, echoing Bernard Stiegler's discussion on philosophy of technology, I pointed at notions of mnemotechnique and mnemotechnology. While the former signifies a manual, attentive, low-scale, slow and controllable mode of exteriorizing and mediating memory, the latter stands for a large-scale system of which automated operations happen way beyond the perception and cognitive capacities of an individual subject whose memories they capture, mediate and process. To exemplify the working of such a mnemotechnological apparatus I used the example of Facebook. While on its surface providing users with a sense of relative control over the way they record and mediate their memories of everyday lives, mnemotechnologies simultaneously capture these memories, organize and process them in order to achieve anticipated commercial benefits. In other words, the architecture of these mainstream mnemotechnologies (its specific design, structure, way of formatting the content, commands encouraging users to share their memories) is today carefully engineered so as to secure the widest accumulation of personal life-bits as possible. Consequently, the rhetoric of empowerment and user-friendliness in discussing everyday life capturing practices taking place via mainstream digital mnemotechnologies needs to become balanced by a recognition of their commercial and techno-economic substructure and what it partly entails: preemptive mechanics. What is meant here is that while capturing our imagination, mainstream mnemotechnologies, increasingly normalized and taken for granted in today's cultures of connectivity (van Dijck 2013), simultaneously capture; in other words, they detain this imagination, and thus preempt possibilities for imagining differently. As I briefly discussed in this chapter, reactions to that state might take various practical forms, sometimes even of a radical nature (for example: counteracting the involuntary capture through practices of data obfuscation).

However, here, as one of the concluding remarks, I would like to point out another potential itinerary in addressing the issue of personal archiving in capture culture, which I have been lately exploring both conceptually and practically through creative media practices (Smolicki 2017). The direction I have in mind here sends us to the concept of the *post-digital* and consequently post-digital memory and archiving practice. Several years before the proliferation of smartphones, network computing and more generally speaking the normalization of pervasive network media in everyday life, the composer and artist Kim Cascone asserted that "the revolutionary period of the digital information age has surely passed. The tendrils of digital technology have in some way touched everyone" (Cascone 2000). He used the term "post-digital" to describe this condition. Florian Cramer explains the post-digital in two ways. First, as an indicator of a period when our fascination with digital systems and gadgets has become historical (we are no longer surprised or fascinated by digital technologies but simply take them for granted). Second, he describes it as a "contemporary disenchantment with digital information systems and media gadgets,"[13] prompted by, for instance, the Snowden revelations.[14] My way of understanding the concept is by seeing it foremost as a critical, reflective and practical revision of the digital. The post-digital is not about declaring the end of the digital and moving on to another phase of technological development, but rather sharpening critical attention towards the current state of digital technologies, their pervasiveness and effects. In reference to the title of this chapter, such a post-digital reflexivity might mean for example an understanding of what happens when we press the button, what motivates practices of capturing, what their consequences beyond our comprehension might be. Echoing Lori Emerson's words expressed in relation to media archaeological modes of inquiry into digital interfaces, the post-digital might mean paying attention to and looking *at* technologies of capture (digital media, their relations, infrastructures) as opposed to passively looking *through* them (2014, 130). In the domain of personal memory and archiving practices which this chapter is primarily concerned with, post-digital might mean two things specifically. On a conceptual level, it might mean a revision of the assumptions related to early enthusiasm revolving around the so-called digital revolution and web 2.0. On a practical level, it might mean a return to non-digital techniques (and hence mnemotechniques) in the way one constructs one's account of everyday life. This return is not about renouncing digital technologies altogether (or counteracting their effects) but rather about establishing productive connections between digital and non-digital means of working with one's memory and (post-) digital legacy. Alessandro Ludovico describes proponents of such a hybrid approach

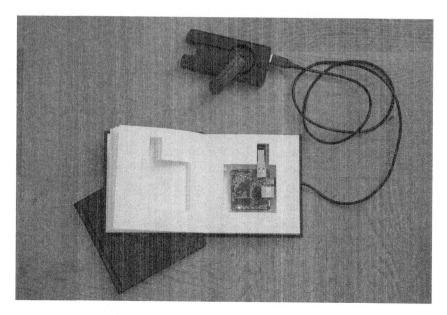

Fig. 1 The author's prototype of a post-digital archive combining a paper note-book and a hacked router with an embedded memory allowing for a controlled upload and sharing of digital memories

to living with media technologies as "neo-analogue media practitioners" and the strategic turn, or as he calls it an "ethical shift," towards non-digital technologies (for instance, handicraft, traditional printing techniques) as a constructive form of resistance to the "go all digital" trend (2012, 155). By implication, this entails a constructive impedance of non-stop surveillance and exposure to mnemotechnological mechanisms that seek benefits in the mass of comparative lumber of digitally mediated memories. To turn this epilogue into an open question: Can the concept of the post-digital inspire new (or regenerate old) creative modes of mnemotechniques forming practical alternatives to the prevalence of mainstream mnemotechnological apparatuses? (Fig. 1).

Acknowledgments I would like to thank my PhD supervisors Maria Hellström Reimer and Ulrika Sjöberg from the School of Arts and Communication at Malmö University. Their generous and inspiring comments and critique helped shape this text, which draws on my recently published doctoral thesis.

Notes

1. Online Etymology Dictionary, accessed January 10, 2015, http://www.etymonline.com
2. Online Etymology Dictionary, accessed January 10, 2015, http://www.etymonline.com/
3. As Wikileaks revealed, the National Security Agency, acting in collaboration with the United Kingdom's M15, conveyed a program called *Weeping Angel* which placed the target TV in a so called "Fake Off" mode while using the internet connection, convincing the owner that the TV is off when it was actually continuously on. In "Fake Off" mode, the TV begins to operate as a bug, recording conversations in the room and sending them over the internet to a covert CIA server. Wikileaks, accessed August 21, 2017, https://wikileaks.org/ciav7p1/cms/page_12353643.html
4. Kit Smith, "Marketing: 47 Facebook Statistics for 2016," *brandwatch*, May 12, 2016. https://www.brandwatch.com/blog/47-facebook-statistics-2016/
5. Mary Lister, "33 Mind-Blogging Instagram Stats & Facts for 2017," *The WordStream Blog*, August 21, 2017. http://www.wordstream.com/blog/ws/2017/04/20/instagram-statistics
6. Craig Smith, "135 Amazing Snapchat Statistics and Facts (September 2017)," *DMR*, October 24, 2017. https://expandedramblings.com/index.php/snapchat-statistics/#.WdX589FdLZt
7. Rebecca Smithers, "Terms and Conditions: Not Reading the Small Print Can Mean Big Problems," *Guardian*, May 11, 2011. https://www.theguardian.com/money/2011/may/11/terms-conditions-small-print-big-problems
8. The full list of datasets of interest to Facebook can be found here: http://europe-v-facebook.org/EN/Data_Pool/data_pool.html. Accessed May 29, 2017.
9. https://labs.rs/en/. Accessed July 2, 2017.
10. Data according to Samuel Gibbs, "Facebook 'tracks all visitors, breaching EU law,'" *Guardian*, March 31, 2015. https://www.theguardian.com/

technology/2015/mar/31/facebook-tracks-all-visitors-breaching-eu-law-report

11. Natasha Singer, "Mapping, and Sharing, the Consumer Genome," *New York Times*, June 16, 2012. http://www.nytimes.com/2012/06/17/technology/acxiom-the-quiet-giant-of-consumer-database-marketing.html?_r=0

12. See Smolicki (2015) for some of my early observations related to the experiment with wearing the device over an extended period of time.

13. From the online article "What is Post-Digital?". http://www.aprja.net/?p=1318. Accessed January 22, 2016.

14. The term "post-digital" has not eluded appropriation by capitalist consumer culture; more specifically, by the design, advertising, and consulting industries. See for instance Lund (2015).

References

Adorno, Theodor, and Max Horkheimer. 1972. *The Dialectic of Enlightenment*. New York: Herder and Herder.

Andrejevic, Mark. 2013. The Infinite Debt of Surveillance in the Digital Economy. In *Media, Surveillance and Identity. Social Perspectives*, ed. André Jansson and Miyase Christensen, 91–108. New York: Peter Lang.

Bell, Gordon, and Jim Gemmell. 2009. *Your Life, Uploaded. The Digital Way to Better Memory, Health, and Productivity*. New York: A Plume Book, Penguin.

Bogost, Ian. 2010. Ian Became a Fan of Marshall McLuhan on Facebook and Suggested You Become a Fan Too. In *Facebook and Philosophy: What's in Your Mind?* ed. D.E. Wittkower, 21–32. Chicago: Open Court.

Brunton, Finn, and Helen Nissenbaum. 2015. *Obfuscation. A User's Guide for Privacy and Protest*. Cambridge, MA: The MIT Press.

Cascone, Kim. 2000. The Aesthetics of Failure: "Post-Digital" Tendencies in Contemporary Computer Music. *Computer Music Journal* 24 (4): 392–398.

Deuze, Mark. 2012. *Media Life*. Cambridge: Polity Press.

Dodge, Martin, and Rob Kitchin. 2007. Outlines of a World Coming into Existence. Pervasive Computing and the Ethics of Forgetting. *Environment and Planning B: Urban Analytics and City Science* 34 (3): 431–445.

Emerson, Lori. 2014. *Reading Writing Interfaces. From the Digital to the Bookbound*. Minneapolis: University of Minnesota Press.

Ernst, Wolfgang. 2012. *Digital Memory and the Archive*. Minneapolis: University of Minnesota Press.

Flusser, Vilém. 2011. *Into the Universe of Technical Images*. Minneapolis: University of Minnesota Press.

Gower, H.D., L.S. Jast, and W.W. Topley. 1916. *The Camera as Historian*. London: Sampson Low, Marston and Co.

Gross, Charles. 1995. Aristotle on the Brain. *The Neuroscientist* 1 (4): 245–250.

Ingold, Tim. 2013. *Making: Anthropology, Archaeology, Art and Architecture*. London: Routledge.

Kember, Sarah. 2014. Ubiquitous Photography and the Ambient Intelligent Amateur. In *Auto. Self-Representation and Digital Photography*, ed. Hans Hedberg, Gunilla Knape, Tyrone Martinsson, and Louise Wolthers. Stockholm: Art and Theory Publishing.

Lovink, Geert. 2011. *Networks Without Cause. A Critique of Social Media*. London: Polity Press.

Ludovico, Alessandro. 2012. *Post-Digital Print: The Mutation of Publishing Since 1894*. Eindhoven: Onomatopee Office and Project-Space.

Lund, Cornelia. 2015. If the Future Is Software-defined, What to Do with Our Hands? Post-Digital Commons and the Unintended from a Design Perspective. In *Post-Digital Culture*, ed. Daniel Kulle, Cornelia Lund, Oliver Schmidt, and David Ziegenhagen. http://www.post-digital-culture.org/clund.

Nafus, Dawn, and Jamie Sherman. 2014. This One Does Not Go Up to 11. The Quantified Self Movement as an Alternative Big Data Practice. In *International Journal of Communication*, vol. 8, 1784–1794.

Schneier, Bruce. 2015. *Data and Goliath. The Hidden Battles to Collect Your Data and Control Your World*. New York and London: W.W Norton & Company.

Smolicki, Jacek. 2015. De-totalizing Capture: On Personal Recording and Archiving Practices. In *ISEA 2015. Proceedings of the 21st International Symposium on Electronic Arts*, Vancouver, Canada, August 14–19, 2015. www.isea2015.org/proceeding/submissions/ISEA2015_submission_153.pdf

———. 2017. *Para-archives: Rethinking Personal Archiving Practices in Capture Culture*. PhD diss., Malmö University.

Stiegler, Bernard. 1998. *Technics and Time, 1. The Fault of Epimetheus*. Stanford, CA: Stanford University Press.

———. 2003. Our Ailing Educational Institutions. *Culture Machine*, 5. http://www.culturemachine.net/index.php/cm/article/view/258/243

———. 2010. Memory. In *Critical Terms for Media Studies*, ed. W.J.T. Mitchell and Mark B.N. Hansen, 64–87. Chicago: University of Chicago Press.

————. 2011. Suffocated Desire, or How the Cultural Industry Destroys the Individual: Contribution to the Theory of Mass Consumption. *Parrhesia* 13: 52–61.

————. 2014. *Symbolic Misery. The Hyperindustrial Epoch.* Cambridge: Polity Press.

————. 2016. *What Makes Life Worth Living: On Pharmacology.* Cambridge: Polity Press.

van Dijck, José. 2007. *Mediated Memories in the Digital Age.* Stanford, CA: Stanford University Press.

————. 2013. *Culture of Connectivity. A Critical History of Social Media.* Oxford: Oxford University Press.

Verbeek, Peter Paul. 2011. *Moralizing Technology, Understanding and Designing the Morality of Things.* Chicago: The University of Chicago Press.

Part II

Consequences of Digital Recording

Ecyclopedias, Hive Minds and Global Brains. A Cognitive Evolutionary Account of Wikipedia

Jos de Mul

1   A Sinner's Confession

Let me begin this chapter with a confession. I am a sinner, too. According to Michael Gorman, former president of the American Library Association, "a professor who encourages the use of Wikipedia is the intellectual equivalent of a dietician who recommends a steady diet of Big Macs with everything" (quoted in Jemielniak 2014a). To make my case worse, I not only encourage my students to use Wikipedia, but I am also guilty of using Wikipedia myself quite frequently. However, I immediately like to add that I hardly ever eat Big Macs, and almost always read and discuss primary texts and reliable secondary literature with my students.

So why do I sin? Well, probably the most obvious reason is the overwhelming amount of information to be found on Wikipedia. At the time of proofreading this chapter (March 28, 2018), the number of articles in

J. de Mul (✉)
Erasmus University Rotterdam, Rotterdam, The Netherlands
e-mail: demul@fwb.eur.nl

© The Author(s) 2018
A. Romele, E. Terrone (eds.), *Towards a Philosophy of Digital Media*,
https://doi.org/10.1007/978-3-319-75759-9_6

the English version alone already had reached 5,599,586, and if we include the articles written in the 298 Wikipedias in other languages, the number exceeded 40 million. Moreover, no other encyclopedia is so up-to-date (the fact that I was able to mention the exact number of articles in the English version of Wikipedia on the day of proofreading was because the Wikipedia lemma on Wikipedia is being updated daily). No wonder that I am not the only sinner: as of February 2014, Wikipedia had 18 billion page views and nearly 500 million unique visitors each month! A second reason I love Wikipedia is the free-access and free-content character of this encyclopedia, offering (worldwide) millions of people, many of them deprived of books and libraries, a wealth of information, knowledge, and sometimes even wisdom.

It seems that Wikipedia even has a divine ring. It promises to provide us with an *omniscience* that once was attributed to God. Together with technologies like telepresence and virtual reality (which express the human desire to obtain two other divine qualities: omnipresence and omnipotence), Wikipedia promises to guide us right through "the pearly gates of cyberspace" (Wertheim 1999).

Why then this guilty feeling when I use Wikipedia? Well, maybe my feeling of sinfulness is not the result of violating my beliefs, but rather of catching myself believing. After all, I do not have a real talent for religion, both in its sacral and profane forms. Being an academic, having digital culture as one of my research subjects, I am well aware of the drawbacks of Wikipedia. Most of them are related to the fact that Wikipedia is a crowd-sourced project, so that every user in principle can be an editor as well. The English version alone has about 140,000 active editors (out of 33 million registered users) and more than 1,200 administrators (editors who have been granted the right to perform special actions like deleting articles, blocking malfunctioning editors and so on). Because of this open character, Wikipedia lacks, according to its critics, accuracy and reliability. There is, for example, no methodological fact-checking. Moreover, there is a rather uneven coverage of topics, the quality of the writing often is poor, and there are many cases of deliberate insertion of false and misleading information and other types of "intellectual vandalism." Other issues that have attracted criticism are explicit content (such as child pornography), sexism, and privacy violations. Indeed, for many reasons the

Wikipedia community resembles a Church. And the list of drawbacks I just gave is far from exhaustive. (For a more complete list—oh irony—it is worthwhile visiting the lengthy and well-documented Wikipedia lemmas "Criticism of Wikipedia" and "Academic studies about Wikipedia.")

2   Encyclopedias and Hive Minds

Instead of giving an exhaustive overview of its drawbacks, in this chapter I want to focus on one essential characteristic of Wikipedia, which has led to much controversy: its collective character. According to enthusiastic supporters of Wikipedia, this project is a paradigmatic example of the so-called Wisdom of the Crowds. According to this idea, popularized by James Surowiecki's book with that title from 2004, group decisions are often better than could have been made by any single member of the group (Surowiecki 2004). As the accuracy of crowd wisdom largely depends on the crowd's size and diversity, it is not surprising, according to Wikipedia supporters, that a survey of Wikipedia, published in *Nature* that same year, which was based on a comparison of 42 science articles with *Encyclopedia Britannica*, found that Wikipedia's level of accuracy closely approached this esteemed encyclopedia (Giles 2005). The diversity of the crowd guarantees the *neutral point of view*, which is one of the key editorial principles of Wikipedia, aiming at a fair and proportional representation of all of the significant views on the topics.

However, more critical minds see Wikipedia rather as a result of the "Stupidity of the Crowds." According to computer scientist and philosopher Jaron Lanier (who, as one of the pioneers of virtual reality, cannot easily be accused of hostility towards digital culture), the most questionable aspects of Wikipedia are what he calls digital Maoism and the Oracle illusion. In *You're not a Gadget*, Lanier argues that, just as in the heydays of the Cultural Revolution in China, in Wikipedia the majority-opinion perspective rules (Lanier 2010). Despite the open character of the Wikipedia project, which, in principle at least, can be joined by everybody and welcomes a variety of opinions, in reality controversial topics often evoke so-called "edit wars," in which particular entries are updated at a rapid-fire rate by representatives of competing political, religious, or ethnic ideologies. According to Dariusz Jemielniak, a long-time editor of Wikipedia and author of the informative book *Common Knowledge? An Ethnography of Wikipedia*, conflict is the dominant mode of interaction in the Wikimedia community (Jemielniak 2014b).[1]

Actually, the majority, having the loudest and most persistent voice, rules. Or rather, we must say, a *particular* majority rules, as Wikipedia exhibits systemic bias, too. The Wikipedia community is heavily dominated by young, male, white, wealthy, English-speaking techno-geeks, whereas women and non-white ethnicities are significantly underrepresented. For example, in a data analysis of Wikipedia lemmas, the Oxford Internet Institute found that only 2.6% of its geo-tagged articles are about Africa, which accounts for 14% of the world's population (Oxford Internet Institute 2012).

Moreover, according to Lanier, "open culture, [especially] web 2.0 designs, like wikis, tend to promote the false idea that there is only one universal truth in [. . .] arenas where that isn't so." As a consequence, Wikipedia's digital Maoism evokes what Lanier terms the "Oracle illusion". As he explains in his book *You're not a Gadget*:

> Wikipedia, for instance, works on what I call the Oracle illusion, in which knowledge of the human authorship of a text is suppressed in order to give the text superhuman validity. Traditional holy books work in precisely the same way and present many of the same problems. This is [one] of the reasons I sometimes think of cybernetic totalist culture as a new religion. The designation is much more than an approximate metaphor, since it includes a new kind of quest for an afterlife. It's so weird to me that Ray Kurzweil wants the global computing cloud to scoop up the contents of our brains so we can live forever in virtual reality. When my friends and I built the first virtual reality machines, the whole point was to make this world more creative, expressive, empathic, and interesting. It was not to escape it. (Lanier 2010)

The phenomenon Lanier refers to is also known as the *hive mind*, as we find it in social insects, whose "mind" rather than being a property of individuals is a "social phenomenon," as it has to be located in the colony rather than in the individual bees (Queller and Strassmann 2009). In a sense, human individuals are social phenomena in the aforementioned meaning, too, as our bodies also are societies of cells that function together to make us walk, clean our blood, digest our food, think, and so on. And even the cells in our body are actually a collection of organelles, or tiny organs, like the energy-producing mitochondria. And if we look at human life on a larger scale, tribes and cities can also be conceived of as superorganisms, displaying hive minds. Thus, both on the micro- and on the macro-level, humans in a very basic way are dividuals rather than individuals. At the macro-level, we could regard the technical infrastructure of Wikipedia as (a tiny part of) an emerging global brain, and the practice of editing and using this global encyclopedia as part of the accompanying "hive mind."

The more so because this technological infrastructure is not "just" a tool for human cooperation. After all, a large part of the authoring and editing of Wikipedia is done by softbots. Besides administrative bots, which perform all kinds of policing tasks, such as blocking spam and detecting vandalism, authoring and editing bots produce a large amount of the content of Wikipedia (Niederer and Van Dijck 2010). Bots are even among the most active Wikipedians, they create a large amount of high-quality revisions and—maybe for that reason—as a group have more rights than registered human users. In 2014, the English version of Wikipedia alone had at least 50,000 bot editors (Jemielniak 2014b) and by March 2018 the most active one, Cydebot, was responsible for no less that 4,534,846 edits.[2] On top of that, when human authors write Wikipedia entries, they are often assisted by administrative and monitoring tools. Rambot, for example, pulls content from public databases and feeds it into Wikipedia, creating and editing uniform articles on specific content, such as cities in the USA and China, while human editors check and complement facts provided by this software robot (Niederer and Van Dijck 2010). From this perspective, the Wikipedia "crowd" is not so much a human crowd, but a hybrid sociotechnical system.

An interesting question is what would inspire a group of multicellular organisms like ants or humans to form a superorganism? In *The Superorganism: The Beauty, Elegance, and Strangeness of Insect Societies*, biologists Hölldobler and Wilson argue that the emergence of superorganisms is a complex process involving genetic evolution and environmental pressures. Generally, a group of insects like honey bees will move from behaving as individuals to forming colonies when they are storing food (like honey or pollen) that comes from multiple sources. At that point, a colony has a better chance of surviving than an individual (Hölldobler and Wilson 2009). This criterion seems to be satisfied by humans as much as by social insects. However, as population genetics suggests, another precondition for the emergence of superorganisms is a strong genetic similarity. This is the case with social insects like ants and bees, and with the cells in our bodies, but is much less the case with human individuals. For that reason, human interaction is characterized by conflict as much as by cooperation. Wikipedia is a good example here: although a paradigmatic example of global cooperation, the aforementioned drawbacks such as edit wars, uneven coverage of content, underrepresentation of women and ethnic groups, the deliberate insertion of false and misleading information show a "generous" variety of conflicts as well. This seems to make the emergence of a real superhuman organism unlikely. As Joan Strassman, an authority on superorganisms, states in an interview: "That [the emergence of a human superorganism] could happen with higher relatedness. But right now there's far too much conflict." She pointed out that the best kinds of cooperative groups for forming an organism are clones, and humans are far from being genetic clones of each other (Newitz 2012).

Moreover, in the case of hybrid sociotechnical systems such as Wikipedia, the relatedness between the idiosyncratic human authors and their mechanical softbot co-authors is even lower. They are not individuals of the same species, and not even part of the same (biological) kingdom. However, as we saw, it is precisely Lanier's fear that Wikipedia is a symptom of a mental "clonization," which threatens to eliminate diversity and individual perspectives. In Lanier's view, Wikipedia functions like the Borg in the *Star Trek* saga, the extraterrestrial species aiming at assimilating all life in the universe, famous for their standard greeting: "We are the Borg. Your biological and technological distinctiveness will be added to our own. Resistance is futile."

Should we consider Wikipedia as the pinnacle of "the wisdom of the crowds" or rather as "the spectre of a Maoist hive mind"? It is clear that at least with regard to this question there is conflict rather than agreement in the discourse on Wikipedia. How to evaluate these sharply contrasting views? In the next section, I will try to find a way out of this controversy by analyzing Wikipedia from the perspective of the cognitive evolution of mankind, focusing on the role computer networks play within this development.

3   Wikipedia in Evolutionary and Historical Perspective

The emergence of information technologies can be regarded as a milestone in the cognitive evolution of mankind, comparable to the two other major transitions of the cognitive structure of the genus *Homo*: the development of spoken language, and the invention of writing. In his book *Origins of the Modern Mind: Three Stages in the Evolution of Culture and Cognition*, the neuropsychologist Merlin Donald gives a fascinating reconstruction of this cognitive evolution of hominids (Donald 1991). Donald distinguishes three stages in this evolution, characterized respectively by a *mimetic*, a *linguistic* and an external *symbolic* cognition.

In his view the highest primates from which man is descended had an *episodic cognition*, that is to say, non-reflexive, concrete, and situation-linked, taking place in a continuous present. At least from *Homo erectus* on, however, a *mimetic cognition* emerged, characterized by the production of conscious, self-initiated representations which were intentional but not (yet) linguistic (see Fig. 1). One might think of the imitation of the behavior of animals and fellow men. According to Donald this evolution had important social implications. Not only did mimetic capability lead to the development of group rituals and bonding in the prehistoric tribes (phenomena that characterize the behavior of small groups right up to the present day), but it also resulted in a great increase in mutual communication and cooperation, as well as in the transfer and conservation of knowledge. This resulted in what we might call a "hive mind light," which probably played an important role in the reproductive success of the genus *Homo* and its distribution all over the globe.

Linguistic cognition made its appearance with *Homo sapiens*. In the course of evolution, the ability to (re)combine basic actions (which, among other things, had developed through the working of stones)

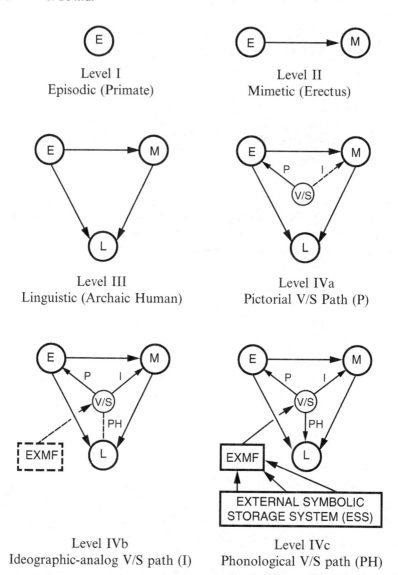

Fig. 1 Four levels. Source: This picture is taken from Donald's *Origins of the Modern Mind* (Donald 1991, 305)

shifted to the production of sound, making articulated language possible. As opposed to mimetic communication, this language makes use of arbitrary symbols, and is characterized by semantic compositionality, which (in comparison to earlier types of animal and human communication), enabled *Homo sapiens* (at least in principle) to utter an infinite number of different sentences with a limited number of words ("John loves Mary," "Mary loves John," "My brother thinks that John loves Mary," "I know that my brother thinks that John loves Mary," "You don't believe that I know that my brother thinks that John loves Mary," and so on.). The implication of this development was another substantial increase of bonding, communication, and cooperation. Moreover, it was also closely connected with the increasing speed of the development, transfer, and conservation of new knowledge and technologies. A hypothesis shared by many biological anthropologists is that linguistic cognition has played a crucial role in the marginalization and eventual extinction of the "archaic" varieties of the *Homo sapiens* (such as the *Homo erectus, denisova,* and *neanderthalensis*).

In the context of Wikipedia, the third transition, from linguistic to symbolic cognition, is of particular importance. "Symbolic cognition" is used here as an umbrella term for cognition mediated by external symbols (Level IV in Fig. 1). Although the origin of this kind of cognition can already be located in archaic *Homo sapiens* (for example, in body painting, markings in bones, which date back at least 300,000 years), symbolic cognition especially emerged in our own species, *Homo sapiens*. From the cave paintings (created *c.*40,000 years ago) onward, in the last 6,000 years it resulted via the icon-based Egyptian hieroglyphs and Chinese ideograms in the modern phonetic alphabet. The invention of writing in particular gave the cultural development of humankind enormous momentum. The power of written culture with respect to the preceding oral culture lies in the fact that it is no longer so difficult to retain and pass on knowledge vital for survival, as such knowledge can now be recorded in an External Symbolic Storage System, duplicated and consulted to an almost unlimited degree.

The transition from linguistic culture to symbolic culture is a fundamental cognitive transformation. This transition implies that one of the key elements of cognition, memory, was transferred (or perhaps more appropriately conveyed in this context as "outsourced") to a non-biological and culturally shared medium. This makes man, to use an expression of biologist and philosopher Plessner, into a species that is artificial by nature (to phrase it differently: from the very beginning being *Homo sapiens sapiens*, has, from the very beginning, been a cyborg: partly organic, partly technological). Furthermore, this transfer implies the emergence of the first material building block of the global brain of a superorganism.

This development had profound cognitive implications: in order to connect the individuals to the External Symbolic Storage System (ESSS), new skills, such as writing and reading, had to be developed (a development made possible by the plasticity of the human neocortex). Liberating thought from the rich but chaotic context of the mainly narrative speech that characterized mythic culture (of which we still find an echo in the written version of originally oral narratives such as the *Odyssey*, the oral encyclopedia of the ancient world [Parry and Parry 1971]), it also allowed thinking to become more precise and abstract. For this reason, Donald says, written culture is profoundly theoretical. Moreover, as has been argued by a number of scholars from the McLuhan school (such as Eric Havelock [1963, 1976, 1986] and Walter Ong [1967, 1982], see De Mul [2010]), along with the medium the message also changed. This can be clearly seen in Plato's philosophy, which according to Havelock reflects the transformation from oral to written culture. Whereas the fleeting oral culture reflects the transient character of everyday reality, due to the fixation, abstraction, and decontextualization brought forth by writing, in Plato's philosophy this transient everyday reality is supplemented by a world of eternal and unchanging Ideas.

Just like the emergence of mimetic skills and speech, symbolic cognition increased the reproductive success of the human species. It played a major role in the transition of hunting and gathering tribes into agricultural societies, characterized by (amongst other things) the emergence of cities, states, social hierarchies, and the revolutionary development of new forms of abstract knowledge and technologies. It also played an important role in social organization. One of the basic problems of agricultural civilizations was the fact that its populations were so huge that the type of personal bonding characteristic for prehistoric tribes was no longer possible. For that reason, as Egyptologist Jan Assmann has argued, it is no coincidence that agricultural civilizations gave rise to monotheist religions, functioning as institutions of social control and repression (Assmann 2006).

Tribal intolerance was replaced by fundamentalist intolerance, characterized by a belief in a universal truth and morality. Writing played an important role in this process, as this "Monotono Theismus" (Nietzsche 1980, VI, 75) is closely connected with holy books. In this sense Jaron Lanier's linking of Wikipedia's ambition to become a kind of universal encyclopedia with the superhuman validity of holy books is not that strange. Or is it? In order to answer that question, we have to take a look at a fourth, more recent cognitive transition: the one from theoretic culture to information culture, characterized by computer-mediated cognition (Table 1).

Merlin Donald published *The Origin of the Modern Mind* in 1991, and although he devotes a couple of pages to the question of what impact the computer might have on human cognition, and even includes a figure to illustrate his point, it is understandable, that (two years before the launch of the world wide web) he wasn't able to dig deep into this impact:

> The very recent combination of this new architecture with electronic media and global computer networks has changed the rules of the game even further. Cognitive architecture has again changed, although the degree of that change will not be known for some time. At the very least, the basic ESS [External Storage System] loop has been supplemented by a

Table 1 Four transitions

Culture	Cognition	Species
First transition: from episodic to **mimetic culture**	Mimetic (rituals)	*Homo ergaster/erectus* (2 million years BC)
Second transition: from mimetic to **mythic culture**	Linguistic (speech)	*Homo sapiens* (100,000 years BC)
Third transition: from mythic to **theoretic culture**	Symbolic (painting, writing)	*Homo sapiens sapiens* (50,000 years BC)
Fourth transition: from theoretic to **information culture**	Computer-mediated (software, networks)	*Homo sapiens sapiens sapiens* (75 years ago)

faster, more efficient memory device, that has externalized some of the search-and-scan operations used by biological memory. The computer extends human cognitive operations into new realms; computers can carry out operations that were not possible within the confines of the old hybrid arrangement between monads and ESS loops shown in the last few figures. For example, the massive statistical and mathematical models and projections routinely run by governments are simply impossible without computers; and so, more ominously, are the synchrony and control of literally millions of monads. Control may still appear to be vested ultimately in the individual, but this may be illusory. In any case, the individual mind has long since ceased to be definable in a meaningful way within its confining biological membrane. (Donald 1991, 358–359)

Now, almost 25 years after the publication of *The Origin of the Modern Mind*, we are in a somewhat better position to point at some fundamental elements of this fourth transition (although I must immediately add that, because we are still at the beginning of this major transition, every analysis of it can only be tentative). I would like to point out three fundamental characteristics, and relate them to Wikipedia and the hive mind.

First, whereas in writing the products of thinking are outsourced to an external memory, in the case of computer-mediated thought, thinking itself (at least those aspects that can be expressed in computer algorithms) is outsourced to an external device. Of course, the computer can also be used as a memory device, but its distinctive character lies in the fact that it functions as an external symbolic processing system (ESPS). This is already the case with the search-and-scan operations Donald mentions in the quoted passage (Google being the most obvious example), but we should also think of expert systems, and (in the big-data age) of all kinds of data mining and profiling.

Second, just as in the case of the transition from oral culture to writing culture, the nature of human thinking transforms. As the outsourcing not only concerns the products of thinking, but the thinking process itself, the result is not so much a fixed, recorded product, but rather a virtual space for thinking, an interactive database which enables the user, assisted by dedicated softbots, to combine and recombine the elements in virtually infinite ways. Computer games offer a good example. They do not so much offer a story (as the novel in writing culture does), but rather a story space, through which the gamer follows his or her own narrative path. In the same way, the scholar in information culture does not present a single argument (such as is the case with the journal article), but rather offers an argumentation space, in which the user creates his own lines of though. One might think of simulations of organisms, economical transactions, or historical events, which enable the user not only to discover all kinds of hidden patterns in the domain of study, but also to make the transition from reality to possibility. Just like the post-mimetic arts, modal natural sciences like artificial physics, synthetic biology, and artificial intelligence aim not so much at a theoretical imitation of reality, but rather at the creation of new realities (Emmeche 1991). But we could also think of a Wittgenstein 2.0 database, in which the user can recombine the thousands of remarks Wittgenstein himself hoped to connect in a flexible network in order to go beyond the fixed character of traditional writing (De Mul 2008).

Third, whereas in Theoretic Culture the connection between individuals and the ESSS is still loose, the connectivity between individuals and the ESPS in the emerging Information Culture becomes increasingly tight. Whereas the traditional mainframe computer was still clearly separated from the users, with the PC, laptop, smartphone, and smartwatch, human beings and computers come ever closer. And if we think of the fast-developing field of implanted computer interfaces, which Google is already experimenting with (Sherwin 2013), we appear to be at the edge of the materialization of the hive mind.

4   Conclusion

Let me return to Wikipedia and conclude with the following realistic observations and speculative predictions (it is up to the reader to decide whether these predictions refer to an informationistic heaven or hell).

Although Wikipedia has so far been quite successful as a collective interactive project, the resulting product in many ways still resembles the products in the ESSS of Theoretic Culture. In spite of its hypertextual form and the outsourcing of all kinds of tasks to softbots, it offers the visitor an encyclopedia that in many respects continues to resemble the encyclopedia projects that have been launched from the Enlightenment on to the *Encyclopedia Britannica.*

As such, Wikipedia also remains prone to the Oracle illusion Lanier (2014) warns us about. Although the continuous process of updating Wikipedia prevents the user from believing in timeless truths, the fact remains that at the moment of visiting only one particular truth prevails. This violates the fundamental principle of information culture that possibility stands higher than actuality. At this moment, so-called "point of view forks" (attempts to evade the neutrality policy by creating a new article about a subject that is already treated in an article) are not permitted in Wikipedia. Although there is a good reason for this policy (avoiding confusion) and for the ambition to collect different viewpoints on an issue in a single article, it also evokes the danger of digital Maoism, and of ending in "the middle of the road" of knowledge.

When I try to imagine the Wikipedia of the future, it will no longer resemble a book, but will rather be a pluralistic argumentation space, an encyclopedic database that will enable the user to traverse and create multiple paths through the subjects, to enter and tweak multimedial simulations of objects and events, and to explore possibilities beyond reality. The user of this *virtual multipedia* will not only think his own thoughts, but (being connected through the global brain with countless other minds) the thoughts of many others as well. However, this hive mind will not necessarily be monotonotheistic, but (if we want) may turn out to be a domain of bewildering hybrid and polytheistic creativity instead.

Acknowledgments This chapter was originally presented at the symposium "Reading Wikipedia," organized by the Dutch Academy of Arts and Sciences on November 23, 2015. The reason for the symposium was that the Wikipedia community was awarded the Erasmus Prize 2015 (http://www.knaw.nl/en/news/calendar/reading-wikipedia?set_language=en). The present version has been updated and adapted to the subject of this book. I would like to thank Alberto Romele for his useful comments on the first version.

Notes

1. Contropedia is an interesting platform for the real-time analysis and visualization of controversies in Wikipedia. http://www.densitydesign.org/research/contropedia-visualizing-controversial-topics-on-wikipedia/. Accessed March 28, 2018.
2. "List of Bots by Number of Edits". *Wikipedia*. https://en.wikipedia.org/wiki/Wikipedia:List_of_bots_by_number_of_edits. Accessed March 28, 2018.

References

Assmann, Jan. 2006. *Monotheismus und die Sprache der Gewalt*. Wien: Picus Verlag.

De Mul, Jos. 2008. Wittgenstein 2.0: Philosophical Reading and Writing after the Mediatic Turn. In *Wittgenstein and Information Theory*, ed. Alois Pichler and Herbert Hrachovec, 157–183. Wien: AWLS.

———. 2010. *Cyberspace Odyssey: Towards a Virtual Ontology and Anthropology*. Newcastle upon Tyne: Cambridge Scholars Publishing.

Donald, Merlin. 1991. *Origins of the Modern Mind: Three Stages in the Evolution of Culture and Cognition*. Cambridge: Harvard University Press.

Emmeche, Claus. 1991. *The Garden in the Machine: The Emerging Science of Artificial Life*. Princeton: Princeton University Press.

Giles, Jim. 2005. Internet Encyclopedias Go Head to Head. *Nature* 438: 900–901.

Havelock, Eric A. 1963. *Preface to Plato*. Cambridge: Harvard University Press.

———. 1976. *Origins of Western Literacy*. Toronto: Ontario Institute for Studies in Education.

———. 1986. *The Muse Learns to Write*. New Haven and London: Yale University Press.

Hölldobler, Bert, and E.O. Wilson. 2009. *The Superorganism: The Beauty, Elegance, and Strangeness of Insect Societies*. New York: W.W. Norton.

Jemielniak, Dariusz. 2014a. Wikipedia, a Professor's Best Friend. *The Chronicle of Higher Education*, October 13, 2014. http://www.chronicle.com/article/Wikipedia-a-Professors-Best/149337

———. 2014b. *Common Knowledge? An Ethnography of Wikipedia*. Stanford: Stanford University Press.

Lanier, Jaron. 2010. *You are Not a Gadget: A Manifesto*. New York: Alfred A. Knopf.

Newitz, Annalee. 2012. Could Humans Evolve into a Giant Hive Mind? *io9. We Come from the Future*, July 3, 2012. http://io9.gizmodo.com/5891143/could-humans-evolve-into-a-giant-hive-mind

Niederer, Sabine, and José van Dijck. 2010. Wisdom of the Crowd or Technicity of Content? Wikipedia as a Sociotechnical System. *New Media & Society* 12 (8): 1368–1387.

Nietzsche, Friederich W. 1980. *Sämtliche Werke: Kritische Studienausgabe*, 15 vols. Berlin: De Gruyter.

Ong, Walter. 1967. *The Presence of the Word*. New Haven and London: Yale University Press.

———. 1982. *Orality and Literacy: The Technologizing of the Word*. London and New York: Methuen.

Oxford Internet Institute. 2012. The Geographically Uneven Coverage of Wikipedia. Accessed March 28, 2018. http://geography.oii.ox.ac.uk/?page=the-geographically-uneven-coverage-of-wikipedia

Parry, Milman, and Adam Parry. 1971. *The Making of Homeric Verse: the Collected Papers of Milman Parry*. Oxford: Clarendon Press.

Queller, David C., and Joan E. Strassmann. 2009. Beyond Society: The Evolution of Organismality. *Philosophical Transactions of the Royal Society B* 364: 3143–3155.

Sherwin, Adam. 2013. Google's Future: Microphones in the Ceiling and Microchips in Your Head. *The Independent*, December 9, 2013.

Surowiecki, James. 2004. *The Wisdom of Crowds: Why the Many are Smarter Than the Few and How Collective Wisdom Shapes Business, Economies, Societies, and Nations*. New York: Doubleday.

Wertheim, Margaret. 1999. *The Pearly Gates of Cyberspace: A History of Space from Dante to the Internet*. New York: W.W. Norton.

Trust, Extended Memories and Social Media

Jacopo Domenicucci

1 Introduction

This exploratory chapter is about the impact of digital technologies on trust. Its aim is to make the case for a non-standard approach to cooperation in a digitally enhanced society. It argues that prominent implications of digital technologies for trust are better understood if we take digital technologies qua means of recording (rather than means of communication). Changes in the memory properties of our environments of interaction are put forward as key to understanding the evolution of trust in a connected society. A research framework is proposed to spell out the impact of recording devices on trust.

We often hear that digital technologies are social technologies.[1] And in fact, there is little doubt about digital devices functioning more and

I am especially indebted to Richard Holton, as usual. I thank Milad Doueihi, Maurizio Ferraris, Laurent Jaffro, Alex Oliver and Enrico Terrone for invaluable discussions. I also thank Pauline Boyer, Bassel Gothaymi, Jens van Klooster, Lamara Leprêtre, Cathy Mason, Victor Parchment and Daphnée Setondji for precious feedback on previous versions.

J. Domenicucci (✉)
University of Cambridge, Cambridge, UK

A. Romele, E. Terrone (eds.), *Towards a Philosophy of Digital Media*,
https://doi.org/10.1007/978-3-319-75759-9_7

more like infrastructures for a variety of forms of cooperation.[2] They allow new forms of shared agency and *update* traditional patterns of social coordination.

What exactly allows digital technologies to have such an impact on our interactions? What makes digital technologies *relevant* for cooperation (and possibly more relevant than other technologies)? We can understand this question along the lines of Herbert Simon's study of the "determinants of the psychological environment of decisions" (Simon 1947, 103). He writes: "human rationality operates, then, within the limits of a psychological environment. This environment imposes on the individual as 'givens' a selection of factors upon which he must base his decisions" (Simon 1947, 117). An ecological approach to decision making is, I take it, what Simon is introducing here.

Digitalization plausibly affects our environments of interaction in a variety of ways. Here I explore the role of a specific factor affecting trust that has been largely overlooked and that, taken seriously, would tell us more about the trust-impact of digital technologies. This factor is memory. Within the contexts of organizations, Simon had already discussed memory as a determinant of this psychological environments of decision: "[. . .] human rationality relies heavily upon the psychological and artificial associational and indexing devices that make the store of memory accessible when it is needed for the making of decisions" (Simon 1947, 99). In this, I explore the idea that, also beyond formal organizations (firms, institutions...), *memory* should be studied as a factor in (mutuating Simon's phrase) our *psychosocial environment of interaction*. And I suggest we take this perspective to look at the evolution of trust in a digitally enhanced society.[3]

The chapter unfolds as follows. The first section suggests the relevance of a memory-centered approach, drawing on some paradigmatic scenarios where digital artefacts play various cooperative roles. The second part brings out the aspects of trust that are affected by memory. The third breaks down "memory" into specific memory-related parameters significant for trust. This twofold analysis should leave us with two sets of features. One distinguishes the different aspects of trust for which memory is relevant. The other tells apart the specific aspects of the memory environment that matter for trust. Together, they are a first step in designing a memory-centered approach to study *the ways in which (digital) recording devices can shape trust*. In the last section, I have a go with part of this resulting framework: I apply it to interpersonal trust powered by Social Media.

2 A Memory-Based Approach

Let us start with typical cases where some form of digitally powered trust is at stake. These are a few among paradigmatic situations you may find yourself in when you live in a digitally enhanced society:

1. *eBay*. I buy a vintage coat on eBay.
2. *Googling*. I google a new colleague I am going to meet.
3. *Friending*. At a party, I friend someone I just met and scroll her Facebook history.
4. *Blockchain*. I buy a flat in Estonia, the first country where real-estate transactions are registered on a digital property ledger fostered by Blockchain technology.
5. *Short Message Service (SMS) browsing*. I check my SMS conversation with a friend to find evidence[4] that we agreed to meet at the Mill and not at the Anchor as he contends.
6. *Border control*. The US border forces have a look at my socials.[5]

These scenarios all involve certain digital tools and are broadly related to cooperative issues. What do these scenarios share beyond the cooperative stakes and their involving a digital device? You may think that in these cases digital technologies are affecting the way people communicate, for example facilitating non-face-to-face interactions. This is part of the story for (1) and (3) but definitely not for (4) and (5). You may think in all these cases the devices are helping me track social reputation features. But this need not be so, as (4) and (5) show. You may think what is common to these scenarios is the role played by some form of artificial intelligence (AI) in powering our exchanges. This is plausible, given the role of sophisticated algorithms in at least (1) to (4). But in (5) the role of AI does not seem to be determinant: this activity is made easier but is structurally indistinguishable from its analogue equivalent (browsing one's correspondence).

I suggest we look at these cases as situations where something is going on with memory. Beyond these situations of loosely construed digital trust, lie different alterations in the memory environment. Part of what makes these interactions possible is tied up with memory properties displayed by the relevant devices. On eBay, you have access to your partner's track record, as evidence of their trustworthiness, and you also know that your own behavior will be recorded and publicly marked and shared, as an incentive to behave. When you google someone you are browsing their public reputation stored online. When you friend someone you gain access to their personal self-presentation and to their social endorsement by their acquaintances. When you register your property on a peer-to-peer distributed ledger, the reliability of the operation is waged upon the virtual impossibility of modifying a stored (and precisely dated) operation (a *block*). When you browse your own older text messages, you draw from a form of extended conversational memory. When the US border agency checks your socials, they in part assess your reliability on the basis of traces you left in a form of social memory. Digital technologies are not just channels enabling telepresent interactions, they are also repositories, ledgers, registries, archives; in a word, extended memories.[6]

There may be different reasons why all these scenarios have *something to do with memory*. This may be due to cooperation itself; maybe memory is just a central variable for cooperation. It may be due to digitally enhanced environments; perhaps being an external memory is central to what it is to be a piece of digital technology. This may be, as seems more plausible, due to both; perhaps digital trust is just at the intersection of the memory needs of cooperation and the memory affordances of technology. Whatever the reasons, the memory properties of our environments of interaction seem to play a role in the way trust is handled by digital technologies. This is the idea I want to illustrate here: a way to understand the digitization of trust is to look at the way digital technologies are impacting the memory properties of our environments of interaction.

It is striking how little discussion of memory you find in both the literature about trust and in the literature about digital trust. There has been, however, a recent turn in *digital studies* towards the idea that digital media are (also) *means of recording*, and not just *means of communication*. Part of my aim here is to draw on this turn to cash out its meaning for digitally enhanced forms of trust. More broadly, and a fortiori, I wish to bring to the attention of trust scholars that *memory matters for trust* (in more than the trivial way in which memory matters to simply any conduct).

It is not easy to fully clarify *why* memory matters to trust, and *why* changes in the memory properties of our environment (like those brought about by the digitization) have an impact on the forms of trust available.[7] This is beyond the scope of this chapter. More relevant for this volume is to put forward *how* trust is affected by the changes in memory properties implemented by the digitization of our environments.

To develop this memory-centered approach, we need to get a grip on two questions. We have to locate the impact of memory on the constituents of trust, and we have to spell out the memory-related parameters that matter. First, for *which* aspects of trust (the reasons, the stakes, the responsiveness . . .) are changes in the memory landscape significant? Second, which properties of the memory situation are determinants of trust, which are the memory-related parameters that matter for trust? I turn to these questions in the next two sections.

A caveat before we enter this discussion. There are two perspectives from which one could discuss the role of digital memories for trust. One is *infrastructural*, the other *conversational*. Here I take the *conversational* approach; that is, I try to focus on the point of view of the participants, partners and interlocutors, and the way digital records change the default memory features of their interactions. My focus is thus on the information participants get about each other, as opposed to the information that they make available to third parties, whether these be the providers of the platforms, others to whom the information is sold, hackers, stalkers or the organs of the state. The *infrastructural* perspective, that I do not explore here, is that of third parties which are the providers (platforms, interfaces and browsers) and the possible intruders (hackers, "doxers" as they are called, state surveillance agencies . . .). If you want, you can look at digital records either standing inside the conversation, or stepping back; you can focus on the mnemonic changes for the agents or for the overall environment. I do the former. Though it is too blunt a distinction, the latter dimension is the one posing the main socio-political puzzles, and it is also the one generally discussed.[8] Diving into the dark side of the records (*dataveillance* ([Clarke 1988], mass surveillance [Feldman 2017] and the "weapons of math destruction" [O'Neil 2016]) is indeed most socio-economically and politically urgent today. But here, focusing on the side of the records which interlocutors have direct access to, I just want to explore what I take to be some ethically relevant effects of this other side of the records. It may well be the tip of the iceberg but it seems significant for the evolution of the communal norms of trust, and for what we expect from each other.

3   To Which Aspects of Trust is Memory Relevant?

Consider what trust would look like in possible worlds deprived of any form of memory, or, conversely, where every single move is committed to eternal remembrance. Would trust even be possible in such worlds? Philosophical research on trust and on digital trust has generally taken memory for granted, overlooking its role as an endogenous factor for trust. Priority is given to the forms of institutions in place, the stakes, the ways in which partners can communicate (namely face-to-face or not). Standard discussions of trust have assumed agents with average mnemonic abilities (neither mnemonic impediments nor artificially powered memories) and environments free of systematic recording (for example, surveillance, distributed ledger transactions . . .). There is no reason to assume so. Just consider how commonly our trusting is affected by our mnemonic limits (Was it this or that cab company that let us down last time?) or our partners' (when someone forgets promises made to us).

Memory is significant for trust in a variety of ways. My memory matters for the ways I can rationally trust (if I cannot remember my past interactions with you, or your reputation, it may be hard to trust you rationally), and for the ways I respond to the trust I am extended (I need to remember by whom and how I am trusted). Assumptions about my partners' memory matter for the way I can trust them (trusting an amnesiac can be a challenge) and for the way I respond to their trust (if I know you tend to forget what you trust me with, I might take it as an incentive to be negligent). Assumptions about third parties' memory can also matter (as when I am aware that I am being audited or monitored). Finally, assumptions about my own memory also matter, or at least the absence of negative assumptions about my mnemonic abilities (if I know that I tend to forget the relevant evidence to trust rationally, I may have to generally revise my trustfulness).

Rather than trying to be exhaustive about the many ways (as you see) in which memory *matters* for trust, let me say a word about three central cases:

1. Memory as a factor for our trusting attitudes. How do the memory properties of our environments of interaction matter for the way we trust our partners?
2. Memory as a factor for trust-responsiveness. How do the memory properties of our environments of interaction matter for the way we respond to the trust we are extended?
3. Memory as a material stake in trust. How can memory-related demands be part of the expectations we have towards our trusted partners?

3.1 Trusting Attitudes and the Evidential Output of Memory

There is little doubt about the importance of being provided with information about our partners (their dispositions, their past, sometimes their intentions . . .) and more broadly about our environment of interaction. It is trivial that various forms of memories are instrumental in providing us with similar information. Under this guise, memory is crucial in allowing us to develop informed, and possibly rational, trusting conducts. This role of memory in trust is supported by experimental results from repeated games settings, for example repeated trust games with patients affected by dementia who showed higher vulnerability than controls (Wong et al. 2017).

That memory is crucial to rational trust does not mean that the relations between trust and memory are linear. Granting discretion to our trusted partner implies that we refrain from putting too much effort into predictions about her behavior, and from constantly assessing her reliability or future behavior. Not only does trust come with a lessened sensitivity to evidence about the other party's possible untrustworthiness, "analogously to blinkered vision" (Jones 1996) (paradigmatically: "I trust her, so I *tend* not to check on her"). It also comes with an implicit non-surveillance clause. For my trusting to be meaningful, I also have the obligation to refrain from checking on my partner, as this potentially limits her freedom (paradigmatically: "I *ought* not to check on my trusted partner"). These factual and normative aspects of the vulnerability of a truster may conflict with certain configurations of the memory properties of an environment of interaction. To keep track of them can be a first step to monitor certain interactions. If I am recording everything, I am not trusting, no matter how much I am relying.

3.2   Responding to Trust, and the Motivational Output of Memory

There is evidence (from Stewart and Plotkin 2016) that in a group where longer memories are developed, partners display more cooperative behavior over repeated games. Stewart and Plotkin conjecture that the correlation between longer memories developed by a group and more cooperative behavior from its members is explained by the fact that improved social records allow more efficient sanctions for defectors. This seems to substantiate the idea that the development of recording systems has a distinctive *motivational output* determinant for the way we react to the trust we are extended. This motivational output seems intertwined with the way memory is a factor in holding one another responsible effectively. This is by the way a core idea in what Nietzsche was announcing at the outset of his *Genealogy of Morality*, when he presented the struggle against natural *forgetfulness* as the logic behind the long history of responsibility and, through responsibility, of moral sociality and cooperation.

If records are often beneficial to trust, they can nevertheless turn out to be detrimental to trust under specific circumstances. Indeed, they may eventually have restrictive effects on our partners' latitude of freedom, comparable to those of surveillance. I could not ask my partners to write down a promise they just made without manifesting that I did not properly take them *on trust*. In similar cases, recording signals diffidence and distrust; and "*Verba volant scripta manent*" may bear some resemblance to "trust is good, control is better."

3.3   Expectations About Memory

As a faculty, a state, an operation, memory can belong to the sort of agency we entrust our partners with. As soon as you accept that mnemonic activities can be understood as involving some form of effort, "trusting someone to remember" is no surprise. The negative side is less obvious, "trusting someone to forget," as it could be understood as some absence of effort (that is, what they are really trusted with is the abstention from making the effort to remember), or some negative effort to forget (which seems empirically substantiated ([Sell 2016]).

Sometimes memory stakes are at the center of an instance of trust, when we have definite expectations about our partners' memory. Sometimes they are background, when they are a condition for further expectations to be fulfilled. The discretion we grant to our trusted partners typically involves memory stakes. If we receive a promise, we tacitly (and minimally) trust our promisers to remember it. We have a variety of expectations about our partners' memory. But expectations about the memory of partners we trust have a distinctive moral dimension. This is twofold. First, we don't just believe that they will remember (remember what they owe us, remember who we are). As it is standardly acknowledged in the philosophical literature on trust, there is a normative dimension to trusting expectations (see, for example, Nickel 2007 and Hawley 2014). Memory seems to be no exception: there are things we think our trustees should remember (which seems to involve taking the relevant measures not to forget). Second, having, as trusters, similar expectations about trustees is also morally relevant sometimes. There are memory-related expectations that we should have if we trust (construed roughly as conditional obligations). Changes in the memory properties of an environment of interaction matter at both normative levels. The available records matter both for the way trustees can fulfill these expectations and for the way trusters' expectations are shaped in the first place.

4   What are the Memory Parameters Significant for Trust?

I think I have given a sense of the relevance of memory for trust. But which properties of the memory situation are precisely the determinants of trust? To operationalize this idea that memory is a variable in trust, we need to break memory into specific memory-related parameters interesting in a situation where memory is not purely mind-based.

From purely mind-based memory to extended forms of memory (that is, a situation where mind-based memories are helped, enriched and sometimes challenged by external records), the main shift, so to speak, is that information is *out there*. This makes information available to a series of operations, such as its management, manipulation, storage and protection that are hardly comparable to their metaphorical mind-based equivalents.

I suggest we distinguish between the management of the information itself and the management of its retrieval. The former is an intervention on the content itself (the information, the data, the message . . .) recorded on a device (or overall recorded in an environment), while the latter is an intervention on the conditions of retrieval of this information. I call the first *amendability*, an art term to designate the question of the possibilities (virtually intrinsic to external records) to modify, alter, enrich and possibly erase information stored on a given record. The second, *accessibility*, frames the question of who has access to given records and how readily. Both are twofold between synchronic and diachronic forms of intervention.[9]

Synchronic accessibility bears on the conditions of accessibility of records at any given time. The conditions of synchronic accessibility determine the level of publicity and privacy for the stored data and information. For example, special conditions of authorization, authentication, identification and clearance are needed to access specific repositories (for example, secret databases but also libraries and e-mail accounts). Who can access the records? Which level of publicity/privacy are they subject to? Which form of authentication or clearance process is necessary for access? What sort of intervention (and by whom) is allowed? Will only the content producers access their content or will third parties (and which) have access?

Synchronic amendability raises questions about who can modify the content and how easily, at a given moment. Can the creator of a content retract it, or correct it, or take it down? What sort of intervention can you have on a message you just sent? How can recipients modify the content they received, recontextualizing it, for example? How easy is it do decontextualize a piece of information?

On the other hand, *diachronic accessibility* is a matter of permanence of these records. How long will they be stored, archived, indexed? The forms of "conservation" define the type and scope of the permanence and durability secured by given records. For example, some records are meant to be permanent (for instance, real-estate public ledgers), others temporary (for instance, a post-it). And *diachronic amendability* is a matter of possible cross-temporal interventions on content. Will content be retractable, modifiable ex post, cancellable? Some records are meant to be unalterable, as with Blockchain distributed ledgers, while others are especially designed to be easily modified by a number of contributors, as with wiki documents.

5   An Example: Social Media and Interpersonal Trust Relationships

Let me exemplify this framework with some remarks on "social media". I use this umbrella term to include both social-media platforms (such as Facebook, Twitter, Instagram, SnapChat, Google+, Flickr, LinkedIn, Pinterest, Tumblr . . .), e-mail services, and instant messaging (SMS, WhatsApp, Viber, Messenger, MSN). This cluster of records matters for a variety of forms of trust, from ecommerce to friendship. Among ICTs, they are the most one-sidedly considered qua means of communications and neglected qua means of recording by philosophers. Beyond their diversity, Social Media share structural features in terms of *accessibility* and *amendability*.

First, *synchronic accessibility*. Digital media allow different levels of privacy and publicity. Overall, however, content tends to be virtually semi-public since it can be easily shared beyond the conversation. It can be easily forwarded, copied, screenshot, and moved from a more private to a more public platform.

Second, *diachronic accessibility*. Most of the content produced, with the exception of very specific platforms (SnapChat) is likely to be stored permanently by the partners we share it with. Storing great amounts of rich data becomes cheap, technically easy and routine on digital compared to analogue means of recording. Hence, digital records may be considered permanent by default. Creators cannot totally erase the data they produced, though they can hide them from their accounts and deindex from search engines (where this applies). And users generally have no clue about how long what they write will last: they lose any control over the life expectancy of their communications.

This takes us, third, to *amendability*, which is very limited for content creators on social platforms. Most messages simply cannot be erased from their partners' device. Rare exceptions can be found, like WhatsApp's, but the possibility to erase from partners' devices (obviously not from the servers) is still generally constrained and comes at its cost, such as a meta-trace indicating that content has been cancelled. While posts can often be taken down, it is impossible to make sure the erased posts do not persist in servers and in partners' screenshots.[10] In this sense, content on social media is potentially committed to eternity.

These effects are decoupled by the technically easy *amendability* for the recipients, which allows decontextualization and recontextualization of digital content. Recorded deeds and utterances can easily gain unexpected and unwanted meaning and can be diverted and exploited. Traces of our digital interactions available to our partners may carry disempowering effects and expose us to new risks often overlooked in ordinary exchange.

Were our relation with "possessors" of our data to take a dramatic turn, they would be in the position to keep, as a hostage, embarrassing moments we try to forget, politically incorrect behaviors we want to hide, or intimate confidences. They would have the basic kit for blackmailing us through informational hostages. The current widely discussed phenomenon of revenge pornography illustrates the dramatic turn that can take the articulation between long term Accessibility, easy synchronic sharing, hardly possible Amendability by the creator. Similar concerns are already clear for the industry. Technical responses go from self-deleting temporary messages (ProtonMail, SnapChat . . .) to platforms reporting to the creator screenshots made on posted content.

In these regards, interactions unfolding through digital media resemble epistolary exchange: they are available to re-reading, their creator cannot renegotiate them, and they can be shared beyond the conversation.[11] Similar concerns are already clear for the industry. Technical responses go from self-deleting temporary messages (ProtonMail, SnapChat...) to platforms reporting to the creator screenshots made on content posted.

We are made responsible by our own accounts[12] and the activity they record, far beyond the way we would explicitly endorse what we do. The track-records constituted by our accounts produce unexpected and often unwanted forms of endorsement. Stored interactions can be compared to some *after the facts contract*,[13] in which all the structuring dimension and the binding effects are ex post and emerge from the (more or less shared) archive. Why should I endorse forever what I once said or did? Why should our partners be in the position to bookkeep our behavior?[14] There is no way to retract what you sent or received. You may manage to hide it, but you cannot properly erase it. Doesn't this situation induce a sense of hyper-accountability comparable to pervasive audit systems?

If social memories structured by Social Media really are comparable to pervasive audit systems, this raises specific worries for trust. The vast literature on the dangers of accountability highlights two ways *systems of accountability* may backfire against cooperation: ill-designed incentives (where audit parameters are satisfied by behaviors actually detrimental to the pursued goals, see the classic (Power 1999)) and crowding out signalling effects (where introducing incentives for a specific behavior seem to erode agents' intrinsic motivations that initially supported the behaviour, see for example (Gneezy and Rustichini 2000)). In the case of Social Media, these risks are all the more perverse that these *systems* are not reliable, since digitally mediated interactions are always pseudonymous, leaving room from pretense. Social Media-related forms of accountability may backfire against trust, without bringing the securing effects of reliable accountable systems. They affect the practical reasoning of honest trustful partner and put her under threat, without deterring the dishonest parties from misleading.

The disempowering effects of social media, due to the forms of *accessibility* and *amendability* they allow, are partially compensated for by a feature of connective sociality, *scalability*, identified by anthropologist Daniel Miller (2016). Users are indeed not passive recipients filling in the gaps offered by these social repositories. There is an active negotiation on the user side, namely allowed by the diversity of platforms available. Their diversity is often considered merely in terms of the various communicative experiences they allow. A crucial trait of *scalability*, though, can be understood from the perspective of recording devices. Accepting certain means of exchange implies accepting this or that standard of recording (duration, publicity ...). These choices have more or less empowering effects on our partners, on how much they can see of us and what they can do with this information.

However, scalability may come at a cost. It looks like we are entrusted with managing our relationships in a more thought-through way than off the record. Sociality has always been scalable, in a loose sense, as we have always cultivated different levels of intimacy, privacy and disclosure with our romantic partners, our friends, our colleagues, our doctors. But this scalability has been made technical by digital platforms, and, hence, explicit and articulated through series of binary thresholds awaiting for a decision. Of course, the struggle for self-presentation is no news. We have always performed on the world's stage. But shifting between different masks requires more self-bureaucracy when we carry detailed semi-permanent and semi-public track-records.

In a sense, the technically scalable connectivity and the policies (norms, conventions and strategies) of social media friending and unfriending seem to head towards a more conscious management of our personal relationships. The sort of skills required to handle social media *with care* may remind us of those of Pirandello's character-type "*il filosofo*", or "*colui che ha capito il giuoco*". These characters typically display a high level of social awareness and a lucid third-person perspective on social order. But these characters also share a common fate of loneliness and unsocial behavior (folly, self-reclusion, identity loss, unsocial behavior . . .). How pro-social is the management of these semi-public semi-permanent records?

For instance, before friending anyone on Facebook, it is reasonable to reflect on how reliable they will be when they tag us on a party picture. To do this is to endorse an objective stance in their regards, to put it in strawsonian terms (Strawson 1962). It implies, to a certain extent, stepping back from the participant stance and evaluating to which extent our partners can be treated as autonomous and responsible. But there are reasons to think that interpersonal trust and the participant stance are strongly intertwined (Domenicucci and Holton 2017). Does *scalability* make explicit similar evaluations and incentivize a managerial (objective) stance towards our partners? Does the fact that platforms are "making sociality technical" (van Dijck 2013) let our trust relationships borrow from the strategies of diplomatic relationships between states?[15] Does this conflict with the background dimension of trust, with its tacitness (Lagerspetz 1998)? Does it potentially carry some sort of Gestalt shift that conflicts with the fact that focusing on trust jeopardizes it?

Now, Amendability and Accessibility clearly are affordances offered by Social Media. Social Media power extremely rich repositories of our interactions, but is it enough to count them as extended social memories? Users' uptake is needed, mere technical possibility is not enough. The tentative argument presented here rests on empirical considerations yet to be explored: do we actually use Social Media as extended social memories? Do we in fact access and amend each other's content on these platforms in ways that are mnemonically relevant, i.e. do we use Social Media beyond instant communication? Let me just note that a positive answer seems rational, psychologically plausible, and technologically likely. Using Social Media as extended social memories seems a rational strategy to reduce uncertainty and vulnerability. It is psychologically plausible that people will actively engage with these repositories, as soon as information costs are sufficiently low (which they already seem to be). Technologically, the basic conditions already seem to be in place (cheap vast memories, Accessibility and Amendability) and further help may come from AI assistants automating certain checking procedures (for example systematically keeping track of the reputation of one's contacts, highlighting kept and unkept promises through semantic analysis of the conversations, and broadly doing background checks—see, for an example of the last one from the industry, *Trooly*[16] and its proprietary system.). This is, after all, a question we can phrase in the terms of the *extended mind* debate. Do Social Media and users qualify as "coupled systems" (Clark and Chalmers 1998)? "Reliable coupling" is the criterion proposed by Clark and Chalmers. It consists in three conditions. First we need Social Media to be a "constant" in users' experience: if available they are consulted. We also need the information they display to be "directly available". Finally, "upon retrieval, information has to be automatically endorsed". It seems very plausible that conditions one and two are realized. The third condition is disputable (though debate on *fake news* shows the complexity of the issue), but extended mind authors already took it to be a more controversial and less stringent condition even in standard cases. Social Media do seem to be more and more akin to *extended social memories*.

6   Conclusion: Pay-Off and Limits of this Approach

I proposed a matrix to frame the investigation of trust in a digitally enhanced society, and I sketched its application to the case of interpersonal trust powered by social media. The ambition of this framework is to be applicable to different forms of trust, from *fides publica* to *fides monetaria* and *fides mercatoria*.

The evolution of the memory environment of our interactions raises questions about its effects on public and political trust. How does trust in public institutions, in political parties and in political leaders evolve when reputation is recorded and capitalized online? How do notions like citizenship and community evolve when (imagined[17]) identities are stored by digital platforms, that is, by private[18] owned social memories?[19] On the *fides mercatoria* side, we are faced with a variety of issues. The development of smart insurance systems, the digitization of firm-customer relationships and the problem of users' control over their personal data, are among the main stakes. Concerning *fides monetaria*, trust in currencies, the study of Blockchain technology, with its distributed ledgers, invites us to reflect on the nature of money (Mallard, Méadel, and Musiani 2014) and on the future of capitalism, if we take seriously the role of records for this form of civilization, as famously highlighted by de Soto (2003).

Besides their local effects on different situations where trust is a stake, widespread digitally recording devices create new bridges and new barriers across traditional modes of interaction. These recording devices carry their own maps of relations: unexpected bridges between friendship and commerce, affection and career, celebrity and information, for-profit and charity, leisure and work. Traditionally distinct registers of trust can become hardly discernible when friendship ends up intertwined with commerce, governed by forms of commercial trust ruled, for example, by a *caveat emptor* principle. The way these evolving maps of sociality affect our very perception of relationships is yet to be studied. Evolutions in the boundaries between personal and impersonal, institutional or peer-to-peer interactions, as well as the privatization of the spaces of encounter, seem crucial stakes here.

Digital trust. Under the heading "digital trust," I tried to advocate a broadening of this research field. As a matter of fact, digital technologies have been mainly investigated as *media*; that is, as means of communication and channels of information.[20] In contrast, this chapter fits a current turn to question this communicational model.[21] Treating digital technologies qua recording devices is all the more relevant for Social Media since research has been mainly concerned with the types of evidence unavailable in online encounters (Pettit 2004; Simpson 2011), the limits to legal devices securing online trust (Hurwitz 2013, 1605–13) and the challenges of telepresence (Daft and Lengel 1986; Lo and Cocking 2008; Lie 2008; Harper 2009; Dreyfus 2015; Simpson 2016).[22]

Memory and trust. Under the heading "memory and trust," I made just a first attempt to bring memory issues to the attention of trust students. After all, this could be a way to renew the take on a question Nietzsche was elaborating in his *Genealogy of Morality*.[23] Indeed, at the outset of the second dissertation, he discusses the forms of mnemonic training necessary to counter natural forgetfulness, and instrumental in the emergence of social morality and cooperation. I am, however, unfaithful to a nietzschean approach, and not only because I gave up on a genealogical method. In this text, Nietzsche clearly stresses the role of memory as a necessary condition for cooperation, taking memory as a cooperation-enhancing ability. While Nietzsche has so seriously warned us of the dangers of memory for our lives (see the *Considerations*, namely the Second), I find no mention of the way under certain conditions memory may threaten specific forms of cooperation. In a sense, here I pursued a complementary line to the one Nietzsche tackled in the *Genealogy*. Memory is indeed crucial for cooperation, but it is not a linear factor nor a single parameter. Specific forms of memory power specific forms of cooperation and inhibit others.

This approach to trust in a digital era has a peculiarity in the way it carves out its object of enquiry. As should be clear by now, I have not discussed "online trust," but trust practices in a digitally enhanced society. Indeed, I have rather looked at trust relations that are cultivated across online and offline contexts, which seems to me the appropriate question for a time where we virtually necessarily conduct a substantial part of our social intercourse through digital media. Discussions of "online trust" are crucial in that they investigate the extent to which we can develop trust in purely computer-mediated interactions, which is relevant for e-commerce, for example. But the large dominance of the field of "digital trust" by research on "online trust" often conveys the idea that the net is a self-contained environment, and issues in "digital trust" are a form of applied philosophy. I contend that reflections on the digitization of trust should be broader than that, and that moral philosophy should care about the evolution of trust relationships in a connected society. Beyond theoretical interests, we cannot step out of the ongoing negotiation about the norms of our trust relationships.

Notes

1. For example, in Tim Berners-Lee's own words: "The web is more a social creation than a technical one. I designed it for a social effect—to help people work together—and not as a technical toy. The ultimate goal of the web is to support and improve our weblike existence in the world. We clump into families, associations and companies. We develop trust

across the miles and distrust around the corner. What we believe, endorse, agree with and depend on is representable and, increasingly, represented in the Web. We all have to ensure that the society we build with the Web is of the sort we intend" (Berners-Lee 2000, 133).

2. See AirBnB, (https://www.airbnb.com/trust, https://www.wired.com/2014/04/trust-in-the-share-economy/) and BlaBlaCar (https://www.blablacar.fr/blablalife/inside-story/in-trust-we-trust) to mention but two among the most popular digitally powered cooperative infrastructures that have emerged in the last few years.

3. Anthropologist Daniel Miller has argued (with evidence from his fieldwork) that it is meaningless to formulate general theses about what digital connectivity does to our relations. For any plausible thesis, we can find an opposed and equally defendable position as soon as we look at another socio-cultural context. Extrapolating slightly, the idea is that digitization has no cross-contextual stable meaning. I contend a memory approach could fit, or accommodate, Miller's razor. This is in part due to the modesty of the proposal, as it does not aim at formulating a single strong thesis about *what the digital does* to trust. This is also due, I believe, to the fact that it tackles a structural feature of digital technologies (their being *recording* devices) which lies behind myriads of their possible uses.

4. This sort of "evidence" can take a dramatic turn, as in the case of this man whose WhatsApp messages were apparently used by his interlocutor to accuse him of blasphemy https://www.theguardian.com/world/2017/sep/16/pakistan-man-sentenced-to-death-for-ridiculing-prophet-muhammad-on-whatsapp

5. See the debate around the current reform of the Department of Homeland Security system of records that intends to "expand the categories of records to include the following: [. . .] the USCIS [United States Citizenship and Immigration Service] Online Account Number; social media handles, aliases, associated identifiable information, and search results." (https://www.federalregister.gov/documents/2017/09/18/2017-19365/privacy-act-of-1974-system-of-records effective from the 18/10/17, document number 82 FR 43556).

6. The present volume is precisely among the first publications (if not *the* first publication) explicitly trying to consider digital media qua recording devices. See Ferraris (2011; 2014; 2015) for a presentation and a theoretical justification of this approach to digital media.

7. This is probably part of Nietzsche's quest at the outset of his *Genealogy of Morals*. I return to that later.

8. See cybersecurity specialist Bruce Schneier on that: "Before computers, what we said disappeared once we'd said it. Neither face-to-face conversations nor telephone conversations were routinely recorded. A permanent communication was something different and special; we called it correspondence. The Internet changed this. [. . .] These conversations [. . .] all leave electronic trails. And while we know this intellectually, we haven't truly internalized it. We still think of conversation as ephemeral, forgetting that we're being recorded and what we say has the permanence of correspondence" ("How to keep your private conversation private for real", Washington Post, March 8, 2017, https://www.washingtonpost.com/posteverything/wp/2017/03/08/conversations-online-are-forever-now-heres-how-to-keep-yours-private/?utm_term=.324abf524c21). See also "The death of ephemeral conversation" (Forbes, October 18, 2016, https://www.schneier.com/blog/archives/2006/10/the_death_of_ep.html) and "The future of ephemeral conversation" (Wall Street Journal, November 21, 2008, https://www.schneier.com/blog/archives/2008/11/the_future_of_e.html).

9. For *accessibility*, this corresponds to two of the main features of archives: "Accessibility" (synchronically) and "Conservation" (diachronically) as defined by Assmann (2012, 329).

10. See the "cofveve tweet" case. The tweet was taken down very quickly by the account that posted it, but this was too late, since many viewers had already screenshot and shared it.

11. Epistolary conversations have always been a laboratory of relationships, where intimacy is built and norms of confidence and privacy are negotiated. For example, Lagerspetz (2014) notes that eighteenth century epistolary culture regarded some letters as belonging to a quasi-public domain, as in Goethe's letters (see his autobiography and his correspondence with Schiller, as well as the narrative usage of letters in *The Apprenticeship of Wilhelm Meister*) which were meant to be read to a group of friends and discussed in a circle: ". . . Goethe tells us that in his youth it was normally expected that private letters might be read aloud in company without anyone thinking it was a breach of trust. In a sense, letters were the blogs of the time. The first volume of his autobiography was published 1811, forty years after the events he describes. At the time, he clearly felt that attitudes had changed so much that he needed to clarify earlier practices to his readers" (Lagerspetz 2014, 141). On the role of letters in the evolution of the sense (and norms) of intimacy, see Eden (2012). Eden studies the rediscovery of Cicero's *sermo familiaris* by Petrarch and then Erasmus

and Montaigne in the perspective of the emergence of the grammar of intimate relationships shaping modern individuality.

12. I can't help noticing that, in their very name, accounts bring their forensic flavor, that we also find in the rest of related terms: "to sign in," "to sign up," "to log in," "to identify," "to authenticate" . . ..

13. I thank Milad Doueihi for this phrase.

14. Even the God of the Old Testament is sometimes asked to forget (and not only to forgive) his creatures' sins, as in *Job* 13: 26: "For thou write bitter things against me and make me to inherit the iniquities of my youth"; and *Psalm of David* 25: "I trust in you [. . .]. Remember not the sins of my youth, nor my transgressions: according to thy mercy remember thou me for thy goodness' sake."

15. Human connectedness and technical connectivity "[. . .] making the Web social" in reality means "making sociality technical." Sociality coded by technology "renders people's activities formal, manageable, and manipulable, enabling platforms to engineer the sociality in people's everyday routines" (van Dijck 2013). While Van Djick, in a provider-centered approach, appropriately focuses on the way the platforms themselves take part in the management of our relationships and networks, here I am highlighting the fact that forms of management or of bureaucracy are found also on the users' side.

16. US patent awarded on June 30, 2015, US9070088B, "Determining Trustworthiness and Compatibility of a Person."

17. Indeed, this area was in a sense pioneered by Benedict Anderson, with his reflection on the role of printing technologies in the widening of political communities and the emergence of nation states (Anderson 1991). If there is historical correlation between print records and national sentiment, which forms of common life are powered by digital records?

18. As noted by (Taeyoon 2017): "However, contemporary society lacks such zones for free association as public spaces are turned into privately owned ones (i.e. community spaces being carefully converted into shopping malls) and mechanisms of surveillance proliferate."

19. Part of these issues is the new research field *fake news* (flourishing in academia since the 2016 US elections), which could also be approached in terms of what widespread recorded conversations do to social epistemology and the public debate. I have tried to develop this perspective in a paper presented at the conference "Web et Post-vérité," March the 9th 2017, Collège d'Etudes Mondiales (Paris), at the international conference "Post-truth, New Realism, and Democracy" at the University of Turin (EHESS, CAS Uniri, Kate Hamburger Kolleg "Recht als Kultur", FMSH),

October the 24th 2017, and at the workshop "La révolution documédiale", Fondation Maison Sciences de l'Homme (Paris), February the 20th 2018.

20. For a review of the literature from this perspective, see, for example, Bargh and McKenna (2004).

21. The present volume is precisely among the first publications (if not *the* first publication) explicitly trying to consider digital media qua recording devices. See Ferraris (2011; 2014; 2015) for a presentation and a theoretical justification of this approach to digital media.

22. That the encounters we can make through digital interfaces do not have the same flavor as face-to-face encounters may be an empirical limit that future forms of telepresence will overcome. In contrast, the high level of control over our self-presentation allowed by digital technologies seems to be a conceptual limit (Cocking 2008, 123–41).

23. "The fact that this problem [for nature to breed an animal able to make promises] has to a great extent been solved must seem all the more astonishing to a person who knows how to appreciate fully the power which works against this promise-making, namely forgetfulness. [. . .] Now, this particular animal, which is necessarily forgetful, in which forgetfulness is present as a force, as a form of strong health, has had an opposing capability bred into it, a memory, with the help of which, in certain cases, forgetfulness will cease to function—that is, for those cases where promises are to be made. ... Precisely that development is the long history of the origin of responsibility" (Nietzsche 2006).

References

Assmann, Aleida. 2012. *Cultural Memory and Western Civilization. Functions, Media, Archives*. New York: Cambridge University Press.

Bargh, John A., and Katelyn Y.A. McKenna. 2004. The Internet and Social Life. *Annual Review of Psychology* 55 (1): 573–590.

Berners-Lee, Tim. 2000. *Weaving the Web: The Original Design and Ultimate Destiny of the World Wide Web*. San Francisco: HarperCollins.

Clark, Andy, and David Chalmers. 1998. The Extended Mind. *Analysis* 58 (1): 7–19.

Clarke, Roger. 1988. Information Technology and Dataveillance. *Communications of the ACM* 31 (5): 498–512.

Cocking, Dean. 2008. Plural Selves and Relational Identity: Intimacy and Privacy Online. In *Information Technology and Moral Philosophy*, ed. Jeroen van den Joven and John Weckert, 123–141. Cambridge: Cambridge University Press.

Daft, Richard L., and Robert H. Lengel. 1986. Organizational Information Requirements, Media Richness and Structural Design. *Management Science* 32 (5): 554–571.

van Dijck, José. 2013. *The Culture of Connectivity. A Critical History of Social Media*. Oxford: Oxford University Press.

Domenicucci, Jacopo, and Richard Holton. 2017. Trust as a Two-Place Relation. In *The Philosophy of Trust*, ed. Paul Faulkner and Thomas Simpson, 149–160. Oxford: Oxford University Press.

Dreyfus, Hubert L. 2015. *On the Internet*. London: Routledge.

Eden, Kathy. 2012. *The Renaissance Rediscovery of Intimacy*. Chicago: The University of Chicago Press.

Feldman, Jessica. 2017. Vérité, confiance et "testaments". Problèmes de privacy et d'écoute éthique dans l'ère de la surveillance de masse ex ante. In *La Confiance à l'ère Numérique*, ed. Jcopo Domenicucci and Milad Doueihi. Paris: Editions Berger-Levrault et Editions Rue d'Ulm.

Ferraris, Maurizio. 2011. *Anima e IPad*. Parma: Guanda.

———. 2014. *Where are You? An Ontology of the Cell Phone*. New York: Fordham University Press.

———. 2015. *Mobilitazione Totale*. Roma-Bari: Laterza.

Gneezy, Uri, and Aldo Rustichini. 2000. A Fine is a Price. *The Journal of Legal Studies* 29 (1): 1–17.

Harper, Richard. 2009. From Tele Presence to Human Absence. The Pragmatic Construction of the Human in Communications Systems Research. In *Proceedings of the 23rd British HCI Group Annual Conference on People and Computers: Celebrating People and Technology*, 73–82. Cambridge, September 1–5, 2009.

Hawley, Katherine. 2014. Trust, Distrust and Commitment. *Noûs* 48 (1): 1–20.

Holton, Richard. 1994. Deciding to Trust, Coming to Believe. *Australasian Journal of Philosophy* 72 (1): 63–76.

Hurwitz, Justin. 2013. Trust and Online Interaction. *University of Pennsylvania Law Review* 161 (6): 1579–1622.

Jones, Karen. 1996. Trust as an Affective Attitude. *Ethics* 107 (1): 4–25.

Lagerspetz, Olli. 1998. *Trust. The Tacit Demand*. London: Springer.

———. 2014. The Worry about Trust. In *Trust, Computing, and Society*, ed. Richard H.R. Harper, 120–143. Cambridge: Cambridge University Press.

Lo, Shao-Kang, and Ting Lie. 2008. Selection of Communication Technologies. A Perspective Based on Information Richness Theory and Trust. *Technovation* 28 (3): 146–153.

Mallard, Alexandre, Cécile Méadel, and Francesca Musiani. 2014. The Paradoxes of Distributed Trust. Peer-to-Peer Architecture and User Confidence in Bitcoin. *Journal of Peer Production* 4. http://peerproduction.net/issues/issue-4-value-and-currency/peer-reviewed-articles/the-paradoxes-of-distributed-trust/

Miller, Daniel. 2016. *How the World Changed Social Media*. http://public.eblib.com/choice/publicfullrecord.aspx?p=4562495

Nickel, Philip J. 2007. Trust and Obligation-Ascription. *Ethical Theory and Moral Practice* 10 (3): 309–319.

Nietzsche, Friedrich. 2006. *On the Genealogy of Morality*. Cambridge: Cambridge University Press.

O'Neil, Cathy. 2016. *Weapons of Math Destruction. How Big Data Increases Inequality and Threatens Democracy*. New York: Crown Publishing Group.

Pettit, Philip. 2004. Trust, Reliance and the Internet. *Analyse & Kritik* 26 (1): 108–121.

Power, Michael. 1999. *The Audit Society: Rituals of Verification*. Oxford and New York: Oxford University Press.

Sell, Andrea J. 2016. Applying the Intentional Forgetting Process to Forgiveness. *Journal of Applied Research in Memory and Cognition* 5 (1): 10–20.

Simon, Herbert A. 1947. *Administrative Behavior. A Study of Decision-Making Processes in Administrative Organization*. New York: Macmillan Co.

Simpson, Thomas. 2011. E-Trust and Reputation. *Ethics and Information Technology* 13 (1): 29–38.

———. 2016. Telepresence and Trust. A Speech-Act Theory of Mediated Communication. *Philosophy & Technology*, online first, September 19, 2016. https://link.springer.com/article/10.1007/s13347-016-0233-3

de Soto, Hernando. 2003. *The Mystery of Capital. Why Capitalism Triumphs in the West and Fails Everywhere Else*. New York: Basic Books.

Stewart, Alexander J., and Joshua B. Plotkin. 2016. Small Groups and Long Memories Promote Cooperation. *Scientific Reports* 6, June, article number 26889.

Strawson, Peter F. 1962. Freedom and Resentment. *Proceedings of the British Academy* 48: 1–25.

Taeyoon, Choi. n.d. *Poetic Computation*. Accessed October 1, 2017. http://poeticcomputation.info/contents/

Wong, Stephanie, Muireann Irish, Claire O'Callaghan, Fiona Kumfor, Greg Savage, John R. Hodges, Olivier Piguet, and Michael Hornberger. 2017. Should I Trust You? Learning and Memory of Social Interactions in Dementia. *Neuropsychologia* 104 (Suppl. C): 157–167.

Digital Eternities

Fanny Georges and Virginie Julliard

1   Introduction

Social networking sites (SNSs) accompany the living throughout their lifetime, providing them with a medium to express their day-to-day thoughts. The technical acculturation highlighted by Josiane Jouët in 1989 regarding the users of the French online videotext system Minitel, and early computers has made headway and deeply embedded technology and the Web in interpersonal relationships and spiritual life. As Jeffrey Sconce showed in his book *Haunted Media* (Sconce 2000), the media have long been intuitively associated with the afterlife. Studies on "techno-spiritual" practices show that the Web is used as a medium or tool for religious practices (Douyère 2011; Jonveaux 2013), be it for consulting

F. Georges (✉)
University of Paris 3 Sorbonne Nouvelle, Paris, France
e-mail: fanny-georges@univ-paris3.fr

V. Julliard
University of Technology of Compiègne, Compiègne, France
e-mail: virginie.julliard@utc.fr

© The Author(s) 2018
A. Romele, E. Terrone (eds.), *Towards a Philosophy of Digital Media*,
https://doi.org/10.1007/978-3-319-75759-9_8

religious content (prayers, sacred texts), for religious observance (for example, locating Mecca for Muslim prayer), religious innovation (for example sending text messages from Jesus) (Bell 2006), or in situations of mourning (Duteil-Ogata 2015). Sites for the practice of "instrumental transcommunication" (spiritism on the Internet) also show that the dead who used the Internet when alive are quite able to express themselves after their death, much like earlier generations' recourse to table-turning (Georges 2013a). For some web users, digital media have become a potential space in which the dead can express themselves.

The social Web has given rise to a phenomenon whereby self-presentation is increasingly delegated to one's community of friends and family and, above all, to the digital writing *dispositif* itself (Georges 2009). Our aim was thus to study the case of a user's death through the prism of how the digital identity the user had created during his lifetime continues to be constructed after his death. At the same time, we observed the actions of the *dispositif* and the signs it produces in this context, along with the ways in which the community of bereaved persons proceed. Since the late 2000s, studies have in fact shown that SNSs (Myspace and particularly Facebook) have become media for expressing grieving (Pène 2011; Wrona 2011; Georges and Julliard 2014). Importantly, the specific features of SNSs remind users of the presence of the deceased among the living, which means the latter are compelled to react to cope with this dysphoric situation. In Facebook, profiles of deceased users remain indirectly active by default. Notifications are regularly sent to their "friends" to encourage them to pick up contact again with the deceased, recommend new friends to them or wish them happy birthday. The *dispositif* can also automatically have invitations sent out from the deceased's account to renew contact or play games played together when the deceased was alive. This form of "persistence" (Brubaker and Hayes 2011) or survival of the deceased may well prolong the distress of their entourage and harbor strong emotional rebounds. As the interviews we conducted show, the fact of seeing the deceased's profile reappear among the living is unbearable for some. Facebook became aware of these issues and since 2009 has given those in the deceased's entourage able to prove the death (death certificate) the possibility to shut down the deceased's pages or to change them into a "memorialized account." In this case, the page displays the word "Remembering" next to the deceased's name and various account functions are disabled: the pages can still be visited by the members of the deceased's social network, but the application no longer sends suggestions to friends encouraging them to interact with him or her, which thus attenuates the traumatic experience of resurgence mentioned by some of the bereaved. For a long time, close family and friends made little use of this feature, either because they were unaware of its existence (Odom et al. 2010) or because people facing this phenomenon are not family members and do not have the necessary administrative papers. In all cases, our interviews showed that it is difficult for bereaved users who had not previously thought about this situation to take decisions or undertake the necessary steps to manage these spaces, which are considered by some as meaningless given the inevitability of death. Users may hesitate to definitively lose all (if the page is deleted) or part (if the page is changed to a memorialized page) of the contents published on their loved one's profile. Yet, we could assume that, given the media's coverage of this subject over the last three years in France, more and more users are now informed and likely to have thought about what they would do in such a situation. Further still, the fact that, since February 2015, users can choose to delete their own profile or have it changed into a "memorialized account" upon death and appoint a legacy contact for their Facebook account will doubtless prompt them to think about their own situation on this count. They can thus choose to keep or delete their data, which avoids their entourage having to make this difficult choice. The service that enables an account to be memorialized is only one of several possibilities for perpetuating the memory of the deceased, but the only one offered by Facebook. Although theoretically not authorized by Facebook, users sometimes prefer to keep the deceased's page "alive," providing they know the login and password, or choose to create a "group" page for remembrance, though these two uses are not mutually exclusive (Georges and Julliard 2014).

In this chapter, we analyze writing practices that change the profile pages of the living into pages of the deceased once they have passed away, whether this involves deleting, modifying or adding content. We accord specific salience to the death announcement, which is a practice that we have chosen to call the "affixment of death stigmas." These changes are carried out by those close to the deceased, who either assume them in their own name or delegate them to a medium or "autonym ligator" (that is, the whole formed by the profile picture and the account name, and understood as being the condensed identity of the web user who owns the account). If the changes are initiated under the "autonym ligator" it is as if the deceased himself were authoring these changes. Complex phenomena of co-enunciation on these pages are then observed: postings by family and friends in their own name exist alongside the deceased's own postings (whether or not these were posted in his or her lifetime) and the signs of survival (due to the account's activity, the deceased seems "persistent and active") that appear alongside death stigmas. The interviewees told us that these changes were designed to bring the deceased's profile into line with the image of him or her that the living wished to keep. We call this phenomenon "profilopraxy." The purpose of this chapter is thus to study how death stigmas are affixed to ensure that the deceased's profile endures, by observing the characteristics of profilopraxy.

The framework we use assumes that this transformation happens immediately after the death, during the first stages of mourning. Kessler and Kübler-Ross (2005) have theorized five successive stages of mourning that currently serve as a reference model: denial, anger, bargaining, depression and acceptance (sublimation). In the first denial stage, the bereaved refuses to accept the death and tries to find elements that bolster this denial. We can thus assume that the deceased's Facebook page can be a kind of artifact that sustains the denial of death thanks to the signs of survival that the bereaved find there. In our analysis, we thus explore the moment that the death is enunciated and link this up with the attempt to deny the death characteristic of the first stage of grieving. The question of enunciating the death on the deceased's Facebook page thus implies investigating the denial of death; however, enunciating the death is surely a means of affixing a death stigma, as we show later, which in fact is at odds with the denial of death.

In this chapter, we present: (1) the overall design of the research methodology; (2) the results relating to the perpetuation of the deceased's profile pages through post-mortem posting activity (the proportion of pages perpetuated and the ways this is achieved); (3) the results concerning the form of enunciations of death, a practice that we term "affixment of the death stigma"; (4) these results enable us to cover some of the practices in profilopraxy.

2   Analyzing Writing Practices that Change the Profiles of the Living into Profiles of the Dead

We approached the process of perpetuating the profiles of deceased persons through the prism of writing practices that intervene after a web user's death (death announcement, expression of sympathy to family and friends, a tribute or address to the deceased), particularly on the SNS profile pages created by users during their lifetime. By the term "writing practices," we mean the production of signs by the user in the space of the techno-publishing *dispositif* (URLs, texts, images, smileys, videos, likes). These publications result from a material inscription and symbolically from an interpretation. Our understanding of the word "writing" is in the broad sense: it includes texts, images, videos, written signs used as graphic symbols. The *dispositifs* of digital *writing* comprise above all media used to symbolically inscribe and organize signs (these are spaces for publishing, documenting and speaking) (Bonaccorsi and Julliard 2010). In the context of death and grieving, writing practices are many: writing practices "in relation to the deceased" include the entourage's production of signs referring to the deceased, whether this involves evoking her within the community of the bereaved or addressing her directly, perhaps revisiting postings that she had published during her lifetime (Georges and Julliard 2016) practices of writing "in place" of the deceased relate to the production of signs by family and friends on the deceased's account. While both practices help to transform the page of a previously living user into the page of a deceased person and affect the representation of the deceased, the second practice especially is a form of profilopraxy, and this for several reasons. First, it intervenes directly on the profile to render it presentable and more reflective of an image of the deceased that corresponds to the one held by the subjects engaged in this practice. Secondly, these changes, like the gestures of thanatopractitioners, are not directly witnessed by the bereaved family and friends. This is particularly the case for deletions and some modifications; unless a longitudinal study is carried out on profile pages ante and post mortem. Moreover, it is not always easy to identify the empirical author of the changes made from within the deceased's account.

We undertook a semio-pragmatic analysis of a corpus of 46 pages created during the users' lifetime, to which we added 37 "group" pages referring to the 43 deceased people studied for the purpose of comparison.

The corpus was built using two procedures. On the one hand, we identified the deceased through the press using the OTMédia database. As a result, most of the deceased whose pages constituted our corpus died following an accident or an illness covered by the press. On the other hand, we identified the deceased by snowballing recruitment (informants close to members of our project). In both cases, we were able to identify the names of the deceased and undertake searches on these names using Facebook's search engine. As not all the names were linked to a Facebook page, we kept only those pages that could be plausibly attributed to the deceased persons, basing this either on the equivalence between the portrait of the deceased given in the press and the description on the Facebook page bearing the deceased's name and on the published content when this was publicly accessible, or by asking for confirmation from our collaborators. In most cases, and perhaps because of the media coverage of the death, we found many "group" pages created post mortem for remembrance purposes. Pages created by the user before he or she died were much more difficult to identify. On the other hand, we had no access to any tributes that may have been produced on the personal pages of the entourage other than through interviews—or when a member of the team was personally involved. (These latter pages were not included in the corpus due to the sensitivity of the data and out of respect for the privacy of this practice, but we did take these into account within an approach that could be described as a participant observation.) To analyze our corpus, we first looked at all of the pages overall and read through the postings. Within this long-term research project, as the contents of these pages involved very sensitive data and were likely to inspire compassion and emotion, we first read these pages in two-person work sessions, which enabled us to share and discuss our impressions. We then carried out an in-depth analysis of the profiles. To respect the right to be forgotten, we made all profiles anonymous. The corpus was analyzed over an 18-month period.

Furthermore, the analysis of the corpus was supplemented by interviews, which enabled us to capture unobservable practices such as the deletion or modification of texts or images. We held seven in-depth open interviews with people from our entourage. Among these, we conducted two interviews over an extended period, taking our inspiration for this from interview practices in the anthropological field, which make it possible to capture the same viewpoint over the different stages of grieving.

3   Perpetuating the Profile Pages of the Deceased

How and in what proportion are the deceased's profile pages perpetuated?

3.1   The Proportion of Perpetuated Pages

The longitudinal study of the corpus of 46 profile pages created by the deceased when they were alive enabled us to divide the pages into four categories (see Table 1). The longitudinal study was conducted as follows: a first screenshot of the pages was taken in September 2014, a second in May 2015. The pages were analyzed at the time of the first and second capture, then verified again six months (October 2015) and one year (April 2016) later. The interviews enabled us to corroborate how the users had proceeded.

Profiles Publicly Active Post Mortem

Out of the 46 pages that we gathered, 23 show public postings post mortem. These pages may also use the "only visible to friends and friends of friends" setting for some postings, or in other words, postings that we

Table 1 Classification of the ways in which deceased persons' pages are perpetuated

Total corpus	46 profile pages
Profiles with public postings post mortem	23
Profiles without public postings post mortem	18
Deleted profiles	4
Profiles transformed into a memorialized page	1

could not necessarily access. Thanks to the snowballing recruitment method, we were able to access more postings than we gathered through the screenshots obtained via the project's Facebook account insofar as the system recognized our personal accounts as "friends of friends" of the deceased person.

Profiles Without Public Postings Post Mortem

These profiles remained untouched, with no mention of the death of the user and no information posted after the user's death. As we only consulted those pages set on "public" mode, our results are relative: out of the 18 pages with no public postings post mortem, some may be active in the "visible to friends only" setting. This figure thus indicates the maximum number of non-active pages in the corpus. In fact, on the pages of the deceased from our entourage (not integrated into the corpus), the post-mortem postings mostly use this private mode, as the users likely consider such information to be too personal for the "public" mode.

Profiles Deleted Post Mortem

Four profiles of deceased persons were deleted between the two retrieval sessions. Among all the names of deceased persons retrieved from the OTMédia database, some may have had a Facebook profile page when alive, and this may even have been changed into a remembrance space but deleted before our first screen-capture (September 2014). For the snowballing recruitment, this information was provided in the interviews. This case thus represents a minimal proportion as, in reality, more pages were probably deleted.

Profiles Transformed into Memorialized Pages

Out of the 46 pages of our corpus, only one has been transformed into a memorialized page.

3.2   Ways of Perpetuating the Page
of a Deceased Person

Half of our corpus shows public postings on the deceased's page, whereas the other half shows none, or these have been deleted. The fact that a page is not publicly updated after a user's death does not necessarily means that it has not been used as a space for remembrance: the page may have been updated by deleting undesirable information, as was mentioned to us in the interviews, or it may have been updated in the "visible to friends only" setting.

Transforming a Living Person's Page into the Page of the Deceased Person: Posting and Remembrance

When a deceased person's page is perpetuated, post-mortem posting activity generally begins with the death announcement. Managing the first reaction to the death on the deceased's page (be it an announcement, tribute, testimony) is a powerful symbolic act. In the case of a tribute or testimony, this act may go further than postings aimed simply at showing one's sympathy or sharing one's pain, which also announce the death indirectly provided they are read.

As the interviews revealed, the deceased person's page is an extremely sensitive medium and subject to a semiotic decision by the bereaved to take on the job of managing it. When the bereaved judge the page to be of little significance, either because the deceased user did not use it regularly or because he published information that in their view does not reflect his personality, the page is not subject to a semiotic decision to manage it (the representation does not appear to family and friends as something to be perpetuated) .

As the decision to delete the page involves laborious procedures with Facebook at a time of grieving, the users—especially if they are not regular Facebook users—will probably consider that the symbolic cost of this step is unjustified owing to the non-representative character of the page: leaving the page untouched thus seems to correspond in some cases to a non-decision. Additionally, the bereaved family and friends who use Facebook regularly can remove the deceased from their Facebook friends to stop receiving re-contact reminders from Facebook, then possibly revisit the deceased's page via a search engine. This configuration may seem more akin to the traditional relationship with the headstone: the bereaved return to the deceased's page and view it as a support for reflection and inner dialogue, recalling through the posts the moment he was still alive. This practice of inner dialogue is not at odds with the use of an updated post-mortem page.

Deletion and Non-Deletion as Enunciative Acts

The practices of writing in relation to the deceased involve not only adding information but also deleting it (deleting information inconsistent with the representation that the user has formed of the deceased person so as to give a better image of the latter, or deleting the profile), or not deleting it (not deleting the deceased's profile or a posting). These are all invisible elements that could only be apprehended through interviews or a meticulous comparison of the different statuses of a profile page at various moments. What is expressed through these acts of deletion or non-deletion is the entourage's wish to perpetuate a certain representation of the deceased and prevent his image from being tarnished by unwelcome postings (for instance, pornographic postings). As a result, the simple fact of leaving the deceased's profile intact may mean constantly monitoring it; it may thus be conjectured that this phenomenon accentuates the feeling of the deceased's presence while the living keep watch over him.

4   Enunciators of the Death and the Places of the Enunciation

When someone dies, the news of their death is passed around somewhat confusedly via word of mouth, by telephone, a written notice, obituary columns or via SNSs. The SNSs modify the temporality of learning that someone has died as they make the news faster or more impromptu. It is not uncommon to learn of someone's death on Facebook even before the family has been informed by the police, for example, after sometimes lengthy procedures; or also to learn of someone's demise long after the date of death, when visiting a friend's page.

4.1   The Announcement of Death via the Techno-Editorial Dispositif

When family or friends wish to transform the deceased's profile page into a memorialized page, it is the techno-publishing *dispositif* that takes over the death announcement text by adding the mention "Remembering" next to the user's name on the cover photo and next to the profile picture. This affixment of the death stigma via the techno-publishing *dispositif*—if we frame the grieving process as described by Kessler and Kübler-Ross [2005]) makes it possible to announce the death while at the same time keeping intact the traces of the deceased's persistence. The bereaved "legacy contact" for the account, who has taken the necessary steps with Facebook to carry out this transformation, may be motivated by the desire to deny the death of the deceased or to carry out the families' desire to deny the death. However, this transformation is irreversible. Once the memorialized page has been created, it is impossible to return to the previous status of the deceased's page. In fact, transforming the profile into a memorialized page is a kind of institutionalization of death, as there is recourse to a third-party authority to announce the death. As a result, this enunciation via the techno-publishing *dispositif* can also be a sign of the acceptance-sublimation of the deceased's death by the mourning legacy contact and/or family mandating him or her—an acceptance-sublimation that marks the end of the grieving process.

4.2   The Death Announced on the Deceased's Page by Family or Friends: the Operator who Affixes the Death Stigma, the Operator who Manages the Page, and the Bereaved Communities

On the deceased's profile page, the death announcement is no longer confined to family, but is open to anyone authorized to publish on the deceased's wall: the traditional divide between family and the entourage, visible in funeral arrangements for instance, has been greatly narrowed. The new role played by friends in perpetuating the deceased's profile and thus his memory does not always function smoothly: for example, when a sick person's death is announced prematurely or if friends inform the family of the death (see above). This is the case, for example, on the page of Alexander, whose death was announced two days before the young man actually died, via condolence messages posted on his "wall," while he lay dying in hospital. However, customs persist and, with time, the family is able to resume its traditional role of maintaining the deceased's page. In fact, in the long run, it is the close family (mother, son, partner or sister) who regularly update the page. Two interviewees talked to us about the creation of a community of "friends" centered around the Facebook page of a deceased person. These communities get together in real life (IRL) in memory of the deceased and may also maintain the page by deleting information or replying to posts. The socialization of grief via the Internet can thus extend to building new sociabilities between the deceased's friends, who were formerly strangers but are now cemented by his death. These communities of friends can support the grieving family by organizing get-togethers in honor of the deceased, for example, or kitties to help to pay for the funeral.

4.3   The Death Announced on the Deceased's Page by the Autonym Ligator

The profile pages of users, after their death, can be managed by family and friends who have the required login and password information even if the death has not been notified via the set Facebook procedure. In this case, the death announcement can be published using the deceased's autonym ligator.

On one page of the corpus where this situation had occurred, to soften the effect of such an announcement (that is, that of the user announcing his own death), the family member identified herself immediately as the enunciator, distinct from the person automatically associated with the autonym ligator: "I'm the daughter of [first name and surname of the deceased] [. . .] for those who don't yet know that my father passed away on 29 August 2014".

On another page of the corpus, the funeral was announced with no mention of the author's name, which produces a curious effect as the deceased himself seems to be inviting his entourage to his own funeral.

We can assume that this form of enunciation makes it possible to comply with the social convention of announcing the death and inviting family and friends to the funeral, while also keeping the deceased quasi-alive insofar as he has seemingly published the announcement of his own funeral. Close family and friends who replied to this posting played along, so to speak, addressing the deceased in the French familiar second-person form "*tu*" (in English "you"): "we will be there for you [deceased's name]." This procedure, which may seem shocking, also enables the bereaved to fully live through the denial stage of grief while maintaining the possibility of dialoguing with the quasi-alive deceased.

On "group" pages created post mortem as a tribute to the deceased, the same effect can be observed: the enunciator identified by the *dispositif* (the deceased) masks the empirical enunciator (the person who created and manages the page). In fact, nowhere is the latter's name mentioned, while the former's name appears everywhere. The profile picture and name are generally those of the deceased: "Tribute to [deceased's first name and surname]." The fact that the creator's name does not appear on the page is sometimes a controversial subject for the deceased's friends and family, who accuse the creator of the page of imposture.

5   The Forms of the Death Announcement and Death Stigmas on the Deceased's Page

Unlike the physical body, which becomes a corpse, the Facebook page bears no direct sign of death produced by the user himself: it is those close to him, referred to as "friends" on the page, who take on the responsibility of transforming the page of the living person into the deceased's profile page. This transformation has consequences, as a personal testimony addressed to the deceased in fact becomes a sort of death stigma (for instance, "damned road" posted as a comment on the last posting that geolocates the departure point for the car of a female user who died in a road accident). This performative act of transforming the Facebook page of a living user into a deceased's profile is especially difficult to manage when the users are not family members. In this section, we differentiate between several procedures for affixing death stigmas on the user's profile.

Death stigmas can be affixed either indirectly or by allusion (meant to be understood by the Facebook "friends" who are already informed), or explicit (for example, "Rest in peace") .

5.1   The Explicit Death Announcement

Death can be announced explicitly on the deceased's page by a relative or friend posting from the deceased's account (see above) or posting a message or comment on the deceased's wall. This announcement explicitly takes different forms that prolong and transform what is observed IRL: funeral announcement, memorial eulogies, conventional formulas.

The Death Announcement Via the Funeral Announcement

More often, the death announcement is communicated through the funeral announcement: "The remains of [deceased's first name] will be laid in the coffin on Tuesday 9 July at the funeral home [place]. The coffin will leave on Thursday between 11 a.m. and 2 p.m. and the service will begin at 2:15 p.m. at the church Saint-Louis. The burial will take place at 3:15 p.m. at the cemetery [town]".

This very factual announcement is no different from a traditional funeral announcement and reproduces the usual text. The funeral announcement is most often posted on the group page created post mortem for memorial purposes rather than on the deceased's own page.

By posting the announcement on the deceased's page, the family can inform people in the entourage who are not in contact with the enunciator of the death and serves in the event that there is no one to relay the information to all of the deceased's network; it enables the *dispositif* to collect the addresses of the entourage, particularly when the entourages of the deceased and the enunciator do not coincide, which is often the case for teenagers and young singles. The very factual nature of this announcement is part of the institutionalization of the deceased's death. It is a formalization, as is the sending of the death announcement, in line with the duty of announcing the death which launches the process of the deceased's social death and grieving. As a result, the traces of enunciation are generally absent from such messages, and instead an impersonal wording predominates as observed on the "group" pages created post mortem in memory of the deceased (Georges 2013b).

The Death Announcement Via a Stereotyped Tribute: the Accepted Formulas

In the pages of our corpus (all by French speakers), the formula "*Repose en paix*" and its acronym RIP (Rest in peace) are frequently employed by users to express their grief and pay tribute on the deceased's page. These expressions are rarely published as main postings: they usually appear as a comment on a message posted by the deceased or to a document representing the deceased (a photo). For example, on one page of our corpus, a female friend posted a comment "RIP [deceased's name]" on July 6, 2013 under the last message posted when the deceased was alive on June 29, 2013; that is, one week beforehand. Here, the expression "RIP" was affixed to the last trace of the deceased's life; following this with her first name, she seems to talk to him in a sort of dialogue with the quasi-living deceased. The affixment of this death stigma triggered an explicit reaction the following day: another female friend posted: "what RIP???" on June 30, 2013. What was an expression of grief had become a death announcement, an affixment of the stigma. In fact, the death announcement was not further formalized, but was followed by other "RIP" postings following on from the first, and over the whole wall. On the other hand, the funeral was clearly announced by a friend on July 14, some 15 days after the death.

One may wonder why the expression "RIP" predominates in a Francophone context. It could easily be interpreted as the manifestation of the extent to which Anglophone and particularly US culture influences Facebook. Moreover, it may also be supposed that, given the discomfort felt by witnesses about mentioning the religious beliefs (Muslim or Christian) undergirding the practices of addressing the dead, in a country like France, which asserts its secularity, the acronym RIP is a more discreet reference to religious tradition than the traditional Christian expression "Rest in peace." The use of "RIP" seems to be more in line with what users consider as an adequate formulation for a medium such as Facebook.

The Death Announcement Via a Remembrance Address

Death can be explicitly announced on the deceased's wall as a remembrance address posted by a friend:

> History will remember that you struggled with all your strength so in your honor we will hold our head high. Being of joy and light, fantastic dancer, magnificent singer, a passionate and endearing friend you will leave all those who knew you with the memory of an unfinished masterpiece. This Saturday May 17 2014 you bowed out. My little brother I send you all my love in the place you are—with [Name of the deceased].

In his remembrance address, Michaël speaks to his deceased friend, Alexander, in the second person. The artistic and social qualities of the deceased and the feelings he inspired in those close to him are mentioned. He is presented as an example to be followed. Death is only alluded to in the paraphrase "bow out," which refers to the deceased's artistic activity. As the message is tagged, it appears on the deceased's page and on the poster's page.

Delegation of the Death Announcement to a Medium

The circulation of texts and the associated techno-semiotic apparatus are characteristic of what is termed the Web 2.0 and its "knowledge circulation economy" (Jeanneret 2014). Facebook, like other SNSs, is set up to deal with the circulation of texts (Julliard 2015). The bereaved can use this possibility to delegate the death announcement to an enunciator outside the community of the deceased's family and friends, by sharing a newspaper article on the deceased's wall, for example, and inviting his or her friends to pass on the information. This is the case, for instance on the page of Thibault, a biker who died in a road accident. The affixment of the death stigma was performed by a female friend who posted on the deceased's wall an article from the website Leprogres.fr titled "After the fatal accident at Sainte-Bénigne," simply accompanied by the mention "share" (November 12, 2013) .

5.2   The Implicit Death Announcement

The death announcement may also be implicit. In our corpus, this type of announcement is mainly seen in comments posted by the deceased's entourage under the messages they have posted, and through profile pictures or cover photographs modified by friends or family members who have the login and password of the deceased's account.

The Death Announcement Via a Comment on a Posting Produced by the Deceased

First, an implicit death announcement can be seen in the comments replying to messages published by deceased individuals during their lifetime and particularly to messages published just before their death.

For example, a female "friend" posted the comment "damned road" on the day that a female Facebook user died. Just before her death in a car accident, the deceased had posted on her wall in public mode geolocation information from a nightclub in Saint-Tropez. Comments then follow that imply knowledge of the death: "courage to you!!! be proud of your princess!"

Announcement Via a Change of the Deceased's Profile Photo

Second, the implicit death announcement may be seen in changes to the deceased's profile picture, generally replaced by a photo showing the deceased with the friend or family member who has taken on the function of legacy contact for the deceased's page. Changing this profile picture for another (which is only possible if the legacy contact has the deceased's login and password) can be understood as being the contact's choice to publish an image that is presumably closer to his or her representation of the deceased. Often, this new image shows the deceased in the company of another person, who we may suppose is the person who made the modification. This is the case, for example, of the profile picture of Edgar, who died on February 21, 2014, and which was replaced twice on August 1, 2014. The profile picture at the time of death was the photo of a horse's head taken at the level of the horse's nostrils, slightly blurred and apparently taken by the teenager. The first post-mortem modification validated by "likes" from 46 people showed a new picture with a young boy in the arms of a woman resembling him and who could be his mother. After a second post-mortem modification, the photo showed the teenager on his horse, galloping, with his face turned to the camera. This photo is still displayed as the profile picture for the account. This modification received 66 "likes" and two comments: "We miss you my brother"; "[Deceased's first name], you remain in our hearts forever." The replacement of the profile picture indicates a change of perspective, as much from the photographic as the enunciative viewpoint: the teenager's photo showing an animal he loved becomes the photo of a parent, probably the mother, who published the image she kept of her child (in her arms during some activity). We can also assume that this replacement was "negotiated": in the end, it is not the intimate family photo that was kept, but the photo of the teenager riding the horse that he obviously held dear, as if the parent who initiated the change of image had had second thoughts about his or her first idea so as to better "pay tribute" to their son. The hypothesis that the choice of a profile picture shows the representation of the deceased most in line with the idea that his entourage has of him can be supported by another example. Although for his first profile picture, Fabien had chosen to publish selfies showing him in a car with a young girl who could have been his girlfriend, the pictures on the "group" remembrance page that his family created when he died show him alone (profile picture) and kissing a woman, possibly his mother, on the cheek (cover photo). While the young man's profile page remains a place for writing and remembrance for some of his entourage, it is above all on the "group" page that most of the post-mortem writing activity appears, an activity that is mainly carried out by his family.

6   Conclusion

In this chapter, we aimed to study how online profiles of the living are transformed into those of the dead. To this end, we first focused on the possibilities available to the bereaved to sustain the deceased's profile and how they take this on. When these sustained profiles are taken in hand, they undergo changes (except in cases where the original pages are left intact so that those grieving can visit them without producing, modifying or removing signs). This transformation is dubbed profilopraxy, whereby the deceased's profile is changed so that it complies with the idea that the bereaved have of the person and/or shows the affixment of death stigmas to make the profile recognizable as that of a dead person. As the most obvious way of affixing these stigmas involves announcing the death of the deceased, we examined this announcement. We identified the enunciators who make the announcement, the places where it appears and the way it is formulated. On the basis of this, we revealed that the characteristics of SNSs profoundly upset traditional hierarchies, since both friends and family intervene on the profile pages and affix death stigmas and shape them for posterity. As a result, the transformation of a living person's profile into a dead person's profile stems from a co-enunciation involving viewpoints that are not always similar. Tensions can even be expressed among the co-enunciators active on a profile. Moreover, some choose to use other spaces in which to produce a representation of the deceased that seems more reflective of the image they wish to see handed down to posterity.

Just as the usage of SNSs has brought about changes in the ways of being together and new modalities of interpersonal communication, we can consider that the way in which people mobilize to pay tribute to the dead goes hand in hand with a change in the relationship to death and to the deceased. Be it the persistence of the profiles, the practices of profilopraxy, or the affixment of death stigmas(which is ultimately a way of feeding the deceased's digital identity), Facebook seems to be the place where death is denied. On Facebook, the dead are part of the world of the living.

References

Bell, Genevieve. 2006. No More SMS From Jesus: Ubicomp, Religion and Techno-Spiritual Practices. *Lecture Notes in Computer Science* 4206: 141–158.
Bonaccorsi, Julia, and Virginie Julliard. 2010. Dispositifs de communication numériques et médiation du politique. Le cas du site web d'Ideal-Eu. In

Usages et enjeux des dispositifs de mediation, ed. Mona Aghababaie, Audrey Bonjour, Adeline Clerc, and Guillaume Rauscher, 65–78. Nancy: Presses Universitaires de Nancy.

Brubaker, Jed R., and Gillian R. Hayes. 2011. We Will Never Forget You [Online]. An Empirical Investigation of *Post mortem* MySpace. In *CSCW '2011 Proceedings of the ACM 2011 Conference on Computer Supported Cooperative Work*, Hangzhou, China, March 19–23, 2011, pp. 123–132.

Douyère, David. 2011. La prière assistée par ordinateur. *Médium* 27: 140–154.

Duteil-Ogata, Fabienne. 2015. New Funeral Practices in Japan. From the Computer-Tomb to the Online Tomb. *Online Heidelberg Journal of Religions on the Internet* 8: 11–27.

Georges, Fanny. 2009. Identité numérique et représentation de soi. Analyse sémiotique et quantitative de l'emprise culturelle du web 2.0. *Réseaux* 154: 165–193.

———. 2013a. Le spiritisme en ligne. La communication numérique avec l'au-delà. *Les cahiers du numérique* 9 (3–4): 211–240.

———. 2013b. *Post-Mortem* Digital Identities and New Memorial Uses of Facebook. The Identity of the Producer of a Memorial Page. *Thanatos* 3 (1): 82–93.

Georges, Fanny, and Virginie Julliard. 2014. Aux frontières de l'identité numérique. Éléments pour une typologie des identités numériques post mortem. In *Les frontières du numérique*, ed. Imad Saleh, Naserddine Bouhaï, and Hakim Hachour, 20–36. Paris: L'Harmattan.

———. 2016. Quand le web inscrit le mort dans la temporalité des vivants. XXe Congrès de la SFSIC, Metz, France, June 8–10, 2016.

Jeanneret, Yves. 2014. *Critique de la trivialité. Les médiations de la communication, enjeux de pouvoir*. Paris: Éditions Non Standard.

Jonveaux, Isabelle. 2013. *Dieu en ligne. Expériences et pratiques religieuses sur Internet*. Paris: Bayard.

Julliard, Virginie. 2015. Les apports de la techno-sémiotique à l'analyse des controverses sur *Twitter*. *Hermès* 73: 191–200.

Kessler, David, and Elisabeth Kubler-Ross. 2005. *On Grief and Grieving*. New York: Simon & Schuster.

Odom, William, Richard Harper, Abigail Sellen, David Kirk, and Richard Banks. 2010. Passing on & Putting to Rest. Understanding Bereavement in the Context of Interactive Technologies. In *CHI 2010. Proceedings of the SIGCHI Conference on Human Factors in Computing Systems*, Atlanta, April 10–15, 2010, 1831–1840.

Pène, Sophie. 2011. Facebook mort ou vif. Deuils intimes et causes communes. *Questions de communication* 19: 91–112.

Sconce, Jeffrey. 2000. *Haunted Media: Electronic Presence From Telegraphy to Television*. Durham, NC: Duke University Press.

Wrona, Adeline. 2011. La vie des morts: jesuismort.com, entre bibliographie et nécrologie. *Questions de communication* 19: 73–90.

Safeguarding Without a Record? The Digital Inventories of Intangible Cultural Heritage

Marta Severo

1   Introduction

Established by the United Nations Educational, Scientific and Cultural Organization (UNESCO) through the Convention of 2003 (UNESCO 2003), the category of intangible cultural heritage is the result of some 30 years of discussion both in the international political arena and in academia. This Convention was designed to create a new protection system for cultural heritage radically different from the traditional system of safeguarding, represented at the international level by the 1954 Hague Convention (for the Protection of Cultural Property in the Event of Armed Conflict) and then by the 1972 World Heritage Convention. While these conventions are meant to protect cultural and natural property of "great importance" (in the definition of 1954) or "exceptional universal value" (in that of 1972), the 2003 Convention seeks to build a democratic system of protection adapted to safeguard all living oral prac-

M. Severo (✉)
University Paris Nanterre, Paris, France
e-mail: msevero@parisnanterre.fr

tices considered to be cultural heritage within the restricted limits of a community. While the 1954 and 1972 Conventions are based on traditional tools of recording and safeguarding, such as the inventory of protected items, drafted by experts, intangible cultural heritage calls for new management tools that can contribute to safeguarding by respecting the living and participatory nature of the practices. From its earliest applications, the 2003 Convention has raised many controversies about how to build inventories without attributing them the permanent and stable character of any form of document (Otlet 1934; Briet 1951; Buckland 1997) and how to reconcile decisional sovereignty of the community with the role of the expert in the production of such inventories. Today, ten years after its implementation, many theoretical and practical aspects remain unresolved. In his book *Warning: The Intangible Heritage in Danger* (2014), Chérif Khaznadar, one of the most important contributors to the content of the 2003 Convention, underlines the paradoxical quality of the intangible heritage category, which, on the one hand, insists on the documentary nature of heritage that should be transmitted as a testimony to posterity, and on the other hand, is characterized by its living, unstable and open nature.

Recently, several observers have drawn attention to the role that digital media could play in solving such a paradox. Today, not only does UNESCO request a video to be published on YouTube as evidence of a practice, but also several national inventory projects (in France, Scotland and Finland) are based on the use of collaborative digital platforms. Considering this situation, this chapter seeks to investigate the contradictory relationship between the inventories of intangible cultural heritage and the concept of document defined as "contents inscribed on fixed and permanent materials in an editorial or a reading context" (Bachimont 2017, 49, our translation). In particular, the objective is to comprehend whether "the digital," through its new forms of production and editing of documents, can solve the puzzle of intangible heritage protection by proposing new ways for recording collective life.

The text is organized in three parts. First, we study the paradox of using the inventory as a system for recording intangible heritage. To do this, we summarize, in the second section, the fundamental features of a traditional cultural heritage protection system by emphasizing the role that documentation, through the construction of lists and inventory files, plays in recording and creating evidence of heritage objects. We also consider the category of intangible cultural heritage by highlighting the elements that make it impossible to use traditional recording tools for its protection. In the third section, we investigate the role played by digital media in proposing a new recording system suitable to intangible heritage. We will rely on the analysis of three digital projects for the inventory of intangible heritage: in Scotland, France and Finland. The objective is neither to carry out a technological or communicational audit of these platforms nor to conduct analysis in the framework of the sociology of technology. By describing the objectives of these projects, we aim to immerse the reader in the issues affecting this sector. Finally, in the fourth section, we return to the concept of document and, in particular, we consider the opposition between document and trace. The objective is to examine the nature and normative power of documents generated by these new digital and collaborative inventory systems. For such a goal, it will be valuable to discuss the distinction proposed by philosopher Maurizio Ferraris (2012) between strong and weak documents, which seems particularly relevant in this context.

2　The Documentary Paradox of Intangible Heritage

2.1　The Inventory as Safeguarding System

The category of cultural heritage, employed in the Middle Ages for labeling private property, was soon extended to include public objects representing the collective identity that ought to be conserved for intergenerational transmission. The concept of cultural heritage involves the idea of a legacy left by generations that precede us, and that we pass on to future generations. If we consider, for example, the definition given by UNESCO, cultural heritage is described as "our legacy from the past, what we live with today, and what we pass on to future generations."[1] Initially, the term "cultural heritage" mainly referred to material objects (sites, historic monuments, works of art, and so on). In recent years, the word has been used well beyond these original contexts, ranging from the extraordinary to the ordinary, from the sacred to the profane, from the material to the ideal, from culture to nature. Despite the massive expansion of these new forms of cultural heritage, the system of safeguarding has not evolved in a comparable way. For a long time and still today, the inventory has been the preferred tool for organizing and recording cultural heritage. Thanks to its capacity to classify and archive information, the inventory, which is nothing other than a type of list, remains the pivot of conservation action. Introduced into the cultural field by curiosity cabinets (Impey and MacGregor 1985), the technique of the inventory has been extraordinarily successful with many institutions, such as museums, archives and libraries. Capable of producing a massive and efficient organization of information, the inventory has become the standard tool for managing collections of material objects. If we consider the inventory as a list that will keep in its memory what cannot be kept in the mind (Leroi-Gourhan 1964), its origins can be traced back to the first writing systems (Goody 1977). It was then rapidly adopted in a variety of contexts to store and classify information. A scientific list not only identifies the characteristics of a phenomenon, but also defines the very nature of the phenomenon by giving it a new form (Latour 1987, 96). Throughout the centuries, inventories have helped to organize species, diseases, books, monuments and knowledge in general. Whether as a catalog, an inventory, a directory, a dictionary or an encyclopedia, the classificatory power of lists shapes knowledge in numerous fields (Bowker and Star 2000).

The inventory has proved to be a particularly suitable tool for cultural heritage. In fact, there is a strong affinity between cultural-heritage objects and inventories: "both depend on selection, both decontextualise their objects from their immediate surroundings and recontextualise them with reference to other things designated or listed" (Hafstein 2009, 93). For this reason, the inventory has played a very important role in the governance of cultural properties, both in the management of information and in the selection of heritage. Regarding information management, the first private collections already had catalogs. Using the model of naturalist collections, the inventory allowed the univocal identification of the property, but also the coherent organization of data, its recording and monitoring of the object over time. Regarding the selection process, no state can disregard the necessity of inventories of national treasures in order to preserve them for future generations (Francioni and Lenzerini 2006, 35). Through inclusion in an inventory, cultural property becomes a document that can be transmitted to future generations. Taking the example of Suzanne Briet (1951, our translation), "the antelope that runs in the African plains cannot be considered as a document [. . .]. But if it is captured [. . .] and becomes an object of study, then it is considered to be a document. It becomes physical proof." According to such a view, the document is a form of recording reality: it "is evidence in support of a fact."

2.2   The Peculiarity of Intangible Cultural Heritage

The origin of the concept of intangible heritage is rooted in the rejection of a safeguarding system based on inventories, lists, classifications and hierarchies, which appears inadequate to protect the cultural heritage of certain countries such as Japan, Bolivia or Peru. In 1984 an initial meeting of experts was organized within UNESCO to establish a program for "non-physical heritage." The Reflection Group succeeded in producing the *Recommendation on the Safeguarding of Traditional Culture and Folklore* in 1989, which evolved over the following years until the 2003 Convention. In the current version of the Convention, intangible cultural heritage[2] is defined as "the practices, representations, expressions, knowledge, skills— as well as the instruments, objects, artefacts and cultural spaces associated therewith—that communities, groups and, in some cases, individuals recognize as part of their cultural heritage" (Article 2 of the Convention). States have a duty to safeguard these practices through various safeguarding actions and mainly through the preparation of national inventories. These inventories have three characteristics that distinguish them from the system described above: (1) they must be drafted by the community (and not by external experts); (2) they are not selective but democratic by including all existing practices (without any selection based on value); (3) they must be living, in contrast to the fixity of the document.

Facing such a situation, Chérif Khaznadar (2014) makes two points. First, according to the author, the Convention, in order to protect a heritage at risk of disappearance, defines a "UNESCO-style" safeguard system based on identification and documentation. Consequently, Khaznadar states: "The convention can become a tool of museification and death" (2014, 28) and do more harm than good to heritage. Secondly, Khaznadar attacks the representative lists. These were created a few years after the validation of the Convention to give visibility to the Convention itself. According to the initial discussion between UNESCO and the States, such lists should include all practices considered to be intangible heritage in order to respect the democratic principle of the Convention. However, because the secretariat could validate only a limited number of applications per year, it was necessary to make a selection, and in order to avoid reproducing the selection methods of the World Heritage Convention, it was decided that practices would be selected on the basis of their exemplarity rather than their outstanding universal value. Yet we find ourselves in this case at the starting point with a selection defined by experts rather than by the "community."

These two criticisms of the Convention raise the general question of whether it is really possible to create a heritage paradigm that is an alternative to the traditional paradigm. As has been said, the latter is based on the transformation of cultural property into a document in order to guarantee its transmission to future generations. Such transformation is carried out mainly through the inclusion of the object in a list (the inventory), an action that would determine the shift from the oral to the written form, from trace to document, and eventually to the death of the living practice. Can inventory be used while respecting the living, informal and consensual quality of intangible heritage? Or is it possible to identify other alternative tools that are not based on the fixed nature of the written document?

The UNESCO Convention proposes an initial answer to this question through the Register of Good Safeguarding Practices. This list does not directly document a cultural practice but the activities intended to safeguard it. Yet this register has not aroused the interest of States and communities and today there are only 17 registered elements compared with the 365 on the Representative List of the Intangible Cultural Heritage of Humanity and 47 on the Urgent Safeguarding List. The activities of promotion and valorization of intangible cultural heritage, essential to support the communities, surely do not impose the crystallization of the practice, yet they do not seem to be able to play the function of safeguard in the way inventories did for centuries. The Convention itself stresses the importance and necessity of inventories: "To ensure identification with a view to safeguarding, each State Party shall draw up, in a manner geared to its own situation, one or more inventories of the intangible cultural heritage present in its territory. These inventories shall be regularly updated" (Article 12). As Chiara Bortolotto (2008) points out in her institutional ethnography of intangible heritage, the pragmatic solution that has been adopted to deal with this "documentary tension" (Bachimont 2017) generated by the 2003 Convention consists in developing an informal level of implementation of the Convention which differs from the formal level of the Convention text. In particular, the author speaks about the "spirit" of the Convention:

> When questioned about the inventories, the 'spirit' of the Convention seems less attached to this process than the text of the Convention. While the text of the Convention requires the creation of exhaustive inventories designed to identify the totality of the intangible heritage items of each country, the secretariat's discourse seems to recognize that exhaustiveness is an unrealistic ambition, while admitting the possible perfectibility of this tool subject to a structural incompleteness. (Bortolotto 2008, 19, our translation)

At this informal level, two elements regain importance: the inventory that can be used as a document while recognizing its perfectible nature; and the expert who can help institutions and communities to find viable safeguarding solutions.

3   The Impact of Digital Media

3.1   Digital Media as Living Environment

Recently, digital media have emerged as a key player in safeguarding heritage. In recent years, more and more institutions have relied on new digital technologies to catalog their collections (Cameron and Robinson 2007). Information and computerization systems have been built to digitize inventories and facilitate their management. These systems not only ensure the sustainability of data, but also contribute to standardizing the heritage selection process (Fraysse 2008). Digital media have also attracted the attention of institutions in charge of intangible cultural heritage. Among them, two types of position can be identified.

Some institutions consider digital media to be the panacea of intangible heritage protection. Surely one of the features of digital media is their power to create a trace. All actions that go through them are voluntarily or involuntarily tracked and recorded. This phenomenon has generated strong enthusiasm both in the commercial world (through the big-data phenomenon) and in research with the explosion of computational social science. The basic idea behind these phenomena is that digital data today constitute a source of information on social life that can enable us to observe *in vivo* facts related to social interactions. According to this viewpoint, in the case of intangible heritage, digital platforms such as YouTube, Wikipedia or Facebook allow community members to leave traces that could be used by the researcher or manager to build "living" inventories. In fact, the inventory would already be there without the need to build it. As an example, Sheenagh Pietrobruno (2013) analyzes the case of YouTube videos of the Mevlevi Sema ceremony. The scholar shows how the videos make it possible to represent the ceremony in a much more participatory way, respecting its evolution over time. Conversely, the inventory file (and also the official UNESCO video) fixes the definition of the practice in a precise temporal moment and with a precise political orientation (in this case, that of the Turkish government which proposed its candidacy). Pietrobruno notes that unofficial YouTube videos show certain ceremonial practices, for example, the fact that today the ceremony can be performed by women in public and dressed in colorful clothes, which is excluded by the official representation in which women can dance only in private and dressed in white. Thus digital traces have two important advantages: they are created by the community and they make it possible to follow the evolution of practices over time. We shall return to this point in the third part of this chapter.

Many other authors have a much more vigilant attitude towards digital media, considering them to be not as useful for safeguarding intangible heritage. According to this second viewpoint, traces available on the internet, such as YouTube videos, are only images of the practices and not the practices themselves. Such images cannot be considered as a suitable way to safeguard living heritage because they are always a way of fixing the practice in a given moment. In fact, they do not really constitute a digital trace but digital data because they have lost the link to their socio-material context of origin (Bachimont 2017).

Rather than arbitrating between these two positions, we prefer to adopt an empirical approach through the analysis of case studies and then to return to the concepts of trace and document. In the next paragraph we will present three projects in which digital media have been used in a similar way to address the documentary paradox of intangible heritage. In Scotland, France and Finland, the national inventory is based on a collaborative digital platform.

3.2 Collaborative Digital Inventories of Intangible Heritage

In 2008, Museums Galleries Scotland, an institution with more than 340 museums and galleries, with the support of the Scottish Arts Council, funded a team of researchers from Edinburgh Napier University to create an inventory of Scottish intangible heritage. Alison and Alistair McCleery, who piloted the project, decided to build a collaborative digital inventory (McCleery et al. 2008; McCleery and McCleery 2016). Given the inadequacy of traditional methods and tools, they chose to build a new platform using tools adapted to the new participatory paradigm. To do this, they implemented two types of initiative: they created a website based on a wiki and organized focus groups to enrich content. "The most appropriate solution—the one that has enabled the twin requirements of accessibility and dynamism to be met—was a customized wiki" (McCleery and McCleery 2016, 191). The fact of providing a deeply democratic (anyone can modify) and living (modifications can be made at any time) digital tool is the element that made this project unique. In its initial version, the online inventory relied on MediaWiki software (http://www.mediawiki.org), which allows users to create an account and add or edit an item. The Scottish wiki could be modified by either authorized or anonymous users. However, modifications by users external to the project were rare and use of the discussion page, which should facilitate exchanges between users, was practically non-existent. After some years, the project was taken over directly by Museums Galleries Scotland. The wiki, with its rather homemade appearance, was replaced by a more professional content management system (Drupal), which attempted to reproduce the democratic beginnings of the original project. Anonymous access was no longer possible and the discussion page was deleted, but the institution relied on a "Wikimedian in Residence" to encourage participation in the wiki and moderate contributions (Orr and Thomas 2016). While maintaining the flexibility of a wiki, this new system organized the content in a more structured way as in an inventory with the presence of fixed fields that had to be completed for each element (such as category[3] and location) and tags that could be added by users (region, time, support, and so on).

In France, Ethnopôle InOc Aquitaine is in charge of the digital inventory of intangible heritage. This institution decided to manage the documentary paradox by opting for two parallel solutions. (1) Migrating the files of the national inventory directly to Wikipedia. The content of each file was published on Wikipedia either by adding to an existing page or by creating a new one. The link to the inventory was indicated by an infobox in the top right. (2) Moreover, the files were published on a dedicated website (www.pci-lab.fr), which has just been put online (October 2017). The strength of this project is its attention to the semantic web. The data are structured in connection with Wikidata (Castéret and Larché 2016). According to the spirit of Wikipedia, participation in the inventory is facilitated through the organization of contributory days (Wikipedia workshops or editathon). Publishing files on the world's largest collaborative platform makes the inventory potentially open to anyone at any time. Moreover, unlike the Scottish project, users do not need to learn the rules of a specific platform; they must simply comply with Wikipedia's method of conduct. However, such organization raises very complex questions about content moderation.

This presupposes, under the aegis of the ministry, a shared governance and the implementation of a collective animation of the tool by favoring the circulation between local and national dynamics for the valorization of intangible heritage. Collaboration is also the outcome of interaction between the website and Wikipedia: the French inventories evolve over time, but only within the framework of a process piloted by the ministry, as the PciLab website does not allow inventory modification. On the other hand, it aims to stimulate contributions to Wikipedia, which already relies on a dynamic community. (Castéret and Larché 2016, 158–59, our translation)

To sum up, the French system is based on a delicate balance between a website that is not open to publishing, and Wikipedia pages that are open to all and where the ministry cannot play the role of moderator. From our viewpoint, this ambivalence reproduces the documentary ambiguity of a traditional inventory, but we shall return to this point in the final part of this chapter.

The third example is the Finnish inventory. This is also based on a collaborative digital platform available online since February 2016. The National Council of Antiquities has chosen a wiki (Wiki-inventory for Living Heritage) based on MediaWiki software. The wiki can be modified only by registered users, but the registration form is open to all. The wiki is believed to animate democratic discussion around candidacies. The National Council of Antiquities serves as moderator and administrator of the platform. It may request changes to the proposed texts or delete inappropriate ones. However, the overall Finnish system for the protection of intangible heritage is not completely open in the way the wiki is. In fact, it relies on the creation of small circles of people linked to each heritage category, and on a group of experts who play a leading role in the heritage selection process. Today the link between these actors and the wiki is unclear.

These examples provide us with two interesting points for our argument. First, in all three cases, the institution in charge of cultural heritage has chosen the internet as the ideal medium for building a more transparent and democratic inventory. In this way, not only does the institution admit the existence of a cause-effect link between the chosen medium and the type of safeguarding action guaranteed by the inventory, but it also admits the existence of a difference between the action of traditional inventories and that of new digital inventories. Second, among digital solutions available, the institution identifies the wiki as the only tool capable of providing accessibility and fluidity to an inventory of intangible heritage that must be living and open to its community. Indeed, the main features that distinguish the wiki from other web-based content-publishing systems are the facts that:

1. It is open to publishing and facilitates the creation of shared knowledge (Aguiton and Cardon 2007). Content can be modified not only by site managers but also by users.
2. It tracks any change.
3. It can provide a discussion page for each page of the wiki. Indeed, the pages of a wiki can be considered a new type of document that does not have the stability of the classical document.

Thus, in all three cases, the choice of a digital platform based on a wiki can be interpreted as a response to the documentary paradox.[4]

4   Between Document and Trace, the Inventory as a Weak Document

Considering the proposed case studies, two theoretical questions merit further attention. First, we aim to investigate the documentary nature of wiki pages and, more generally, the ability of the internet to create new types of document, or better signs, that do not have the fixity of the document in its classic definition (Otlet 1934; Briet 1951; Buckland 1997). To do so, it becomes worthwhile to review the distinction between digital document and digital trace, and to consider on which side of the scale contributions to a wiki go. Second, if we can recognize the existence of a new type of document, it will be necessary to question its normative power. In other words, can these new wiki-inventories be as effective in safeguarding cultural heritage as traditional inventories?

Considering the limits of this chapter, we would like to focus our attention on the contribution of Italian philosopher Maurizio Ferraris on this issue. In his theory of documentality (2012), a document is a social object, an "inscribed act." Social reality is based on documents. Using Derrida against Searle, and Searle against Derrida, he attributes an ontological priority, but just in the social world, to registration, inscription, and writing over communication and orality. He also introduces the "documentary pyramid," where documents are situated above three other layers:

1. The trace, which is the basis of the pyramid, is a sign that has been generated by events, without signification or intention.
2. The trace becomes the registration when it is generated by a support that is designed to preserve it over time: for example, the recordings of a camera or a microphone, but also something memorized passively in the brain. Registrations must be accessible to at least one person.
3. Then, registration becomes inscription when its knowledge is intentionally shared with at least two people and becomes a social fact. The intention to leave a sign differentiates inscriptions from recordings.
4. Finally, inscription becomes document when the trace obtains an institutional form, and it is precisely the institutionalization that leads to the fixation of the trace.

Another important element, for our purposes, is the distinction Ferraris proposes between weak and strong documents. A document in the strong sense (the legal document) is the inscription of an act. A document in the weak sense is the recording of a fact. Weak and strong documents both have social value. Yet, the strong document also has normative, institutional and political value. From an ontological point of view, the strong document is the inscription of an act having its own agency, where the weak document is only proof. The author explains: "In this scheme, a document in the strong sense is mostly linked to writing, while one in the weak sense may be, as in the case of traces and discoveries, connected rather to archiwriting" (Ferraris 2012, 267). Archiwriting is writing around writing; it embraces "the thousands of ways we keep track of everyday experience and the world around us" (Ferraris 2012, 207): rituality, memory, animal traces, and so on.

If we return to consider our case studies, we can develop two points. First, the distinction between *digital trace* and *digital document* is less relevant than expected. The contribution published on a wiki is the result of the passage from trace to registration (the platform makes it possible to record any change) and then from registration to inscription (the wiki is a collective and intentional system). This allows us to explain why cultural heritage's institutions are so attracted to participatory digital platforms: they create inscriptions that do not have the fixedness of the document but which, at the same time, maintain social value and permanence over time well above the trace.[5] Second, the distinction between *strong* and *weak document* allows us to question the relationship between these wikis and their institutional value and, consequently, their safeguarding power. The concept of weak document is particularly appropriate for describing the characteristics of the wiki: its dynamism and openness, as well as its weakness with regard to its social and institutional action.

These are two sides of the same phenomenon. It can be effectively observed in the three case studies that we propose: in each one, the wiki is proposed as a solution to avoiding the fixity of the document; but, at the same time, this solution always loses institutional power. In all three cases, the institution seems to "feel" this tension between weak and strong documents, between a social force and normative weakness, and in all three cases, seeks to overcome the obstacle with idiosyncratic solutions. For this reason, the first Scottish wiki, which was completely open and based on MediaWiki, was replaced by a "fake" wiki, which was controlled and easier to moderate. And this is also the reason why in France the safeguarding system is based on a separate website that contains the official inventory and is not open to publication. Similarly, in Finland, the wiki favors discussion around candidacies, whereas the institutionalization of an element is always established outside the digital platform.

Considering all this, these new platforms do not seem to constitute a definitive solution to the documentary paradox. They are, rather, a workaround and, at the same time, display a desire for openness and transparency that is not really achievable in today's cultural-heritage system.

Yet, on the basis of Ferraris's book *Mobilitazione Totale* (2015),[6] we can take this reflection even further. In this text, the philosopher applies his theory of documentality to contemporary society, dominated by new digital technologies. According to him, the use of digital technologies always corresponds to a call to action that has normative value. Digital devices that go by the acronym ARMI (which stands for "*Apparecchi di Registrazione e Mobilitazione dell'Intenzionalità*," "Devices for the Registration and

Mobilization of Intentionality," but which also means "weapons" in Italian) generate a system of "total recording" that leads to the total mobilization of human beings. The internet is seen as an accelerator of documentality that makes recording and registration limitless. All that is online is then a document. But is it a strong or a weak document? In *Mobilitazione Totale*, Ferraris redefines this distinction in these terms: "The strong document is the document that owns a power [. . .]. The weak document is the document that only keeps track of what happened [. . .]. These documents have a simple informational and not normative power" (2015, Chap. 3 *Il potenziamento tecnologico: la rete*, our translation). A little further on, he states: "the web is a performative system, not merely descriptive system." He therefore recognizes in web documents a social force even if they do not have an explicit normative value.

While the opposition between social force and the "force of law"[7] is not new, what is interesting is the fragility of the border separating these two. As Wittgenstein (1986) points out in relation to language, while we can distinguish between the rules of language, on the one hand, and everyday use on the other, items of everyday use are quite often integrated into the normative system of language and become essential for its comprehension.

Similarly, we do not know exactly what the normative effects of digital inventories of intangible heritage will be in the future. Today, they fall into the category of weak documents, compared with the traditional legal system of safeguarding. Nevertheless, digital technologies constitute a power system, which, through its affordances, could put into crisis the distinction between weak and strong document and thus facilitate the transformation of the social force of inventories into the force of law. Yet, in this case, would it be only a question of time (the time needed to recognize the authority and normative value of a wiki, in this case) or rather one of the manifestations of the conflict, always unresolved, between the social and the institutional?

Acknowledgments This work was supported by the Labex *Les passés dans le présent* (Investissements d'avenir, réf. ANR-11-LABX-0026-01).

Notes

1. "World Heritage," UNESCO, accessed September 15, 2017, http://whc.unesco.org/en/about/
2. The choice of such a denomination is due principally to the need to avoid the word "folklore," which in the European context and especially in France would have a racist meaning (following the use of the term during the war).
3. The elements were then divided into 12 categories (compared with the five official categories of the Convention), which also included practices such as games and culinary traditions.
4. Before passing to the fourth section, it is important to recognize the difficulties that these projects have encountered and are still encountering in soliciting the participation of community members. In particular, cultural heritage managers have to cope with the digital divide related to age, lack of computer skills or simply to a reluctance towards digital media. However, this question is irrelevant to the purpose of this chapter, which questions the theoretical relationship between intangible heritage and documents without going into the problems of a practical implementation.
5. We do not reject the interest of the concept of digital trace, which in other contexts we have used extensively (Severo and Romele 2015). Yet, in this context, the absence or weakness of intentionality of traces makes them irrelevant for the construction of inventories where recognition by the community constitutes a determining element for safeguarding.
6. A previous and shorter version of the thesis contained in this book is available in English in Ferraris (2014).
7. Jacques Deridda (1990) in particular investigates the difference between force of law and justice: a distinction that we can recognize implicitly in the opposition drawn by Ferraris between strong and weak document.

References

Aguiton, Christophe, and Dominique Cardon. 2007. The Strength of Weak Cooperation. An Attempt to Understand the Meaning of Web2.0. *Communications & Strategies* 65: 51–65.
Bachimont, Bruno. 2017. *Patrimoine et numérique. Technique et politique de la mémoire*. Paris: INA Editions.

Bortolotto, Chiara. 2008. *Les inventaires du patrimoine culturel immatériel: l'enjeu de la participation.* Rapport de recherche, Direction de l'architecture et du patrimoine.

Bowker, Geoffrey C., and Susan Leigh Star. 2000. *Sorting Things Out. Classification and Its Consequences.* Cambridge, MA: The MIT Press.

Briet, Suzanne. 1951. *Qu'est-ce que la documentation?* Paris: Éditions documentaires, industrielles et techniques.

Buckland, Michael. 1997. What is a Document? *Journal of the American Society for Information Science (1986–1998)* 48 (9): 804–809.

Cameron, Fiona, and Helena Robinson. 2007. Digital Knowledgescapes. Cultural, Theoretical, Practical, and Usage Issues Facing Museum Collection Databases in a Digital Epoch. In *Theorizing Digital Cultural Heritage. A Critical Discourse*, ed. Fiona Cameron and Sarah Kenderdine, 165–192. Cambridge, MA: The MIT Press.

Castéret, Jean-Jasques, and Mélanie Larché. 2016. Le projet 'PciLab' pour la valorisation numérique de l'Inventaire français du PCI. In *Patrimoine culturel immatériel et numérique*, ed. Marta Severo and Séverine Cachat, 147–162. Paris: L'Harmattan.

Derrida, Jacques. 1990. *Du droit à la philosophie.* Paris: Editions Galilée.

Ferraris, Maurizio. 2012. *Documentality. Why It is Necessary to Leave Traces.* New York: Fordham University Press.

———. 2014. Total Mobilization. Recording, Documentality, Normativity. *The Monist* 97 (2): 201–222.

———. 2015. *Mobilitazione Totale.* Roma-Bari: Laterza. Kindle.

Francioni, Francesco, and Federico Lenzerini. 2006. The Obligation to Prevent and Avoid Destruction of Cultural Heritage: From Biniyan to Iraq. In *Art and Cultural Heritage. Law, Policy, and Practice*, ed. Barbara T. Hoffman, 28–41. Cambridge: Cambridge University Press.

Fraysse, Patrick. 2008. Effets du système d'information sur l'évolution de la notion de patrimoine. In *L'information dans les organisations. dynamique et complexité*, ed. Christiane Volant, 303–314. Tours: Presses Universitaires François-Rabelais.

Goody, Jack. 1977. *The Domestication of the Savage Mind.* Cambridge: Cambridge University Press.

Hafstein, Valdimar. 2009. Intangible Heritage as a List. From Masterpieces to Representation. In *Intangible Heritage*, ed. Laurajane Smith and Natsuko Akagawa, 93–111. London: Routledge.

Impey, Oliver, and Arthur MacGregor. 1985. *The Origins of Museums. The Cabinet of Curiosities in Sixteenth- and Seventeenth-Century Europe.* Oxford: Clarendon Press.

Khaznadar, Chérif. 2014. *Warning. The Intagible Heritage in Danger.* Arles: Actes Sud.

Latour, Bruno. 1987. *Science in Action. How to Follow Scientists and Engineers Through Society.* Milton Keynes: Open University Press.

Leroi-Gourhan, André. 1964. *Le geste et la parole.* Paris: Albin Michel.

McCleery, Alistair, and Alison McCleery. 2016. Inventorying Intangible Heritage: The Approach in Scotland. In *Patrimoine culturel immatériel et numérique,* ed. Marta Severo and Séverine Cachat, 183–198. Paris: L'Harmattan.

McCleery, Alison, Alistar McCleery, and Linda Gunn. 2008. *Scoping and Mapping Intangible Cultural Heritage in Scotland. Final Report.* Edinburgh: Napier University and Museums Galleries Scotland. http://www.napier.ac.uk/~/media/worktribe/output-229389/ichinscotlandfullreportjuly08pdf.pdf

Orr, Joanne, and Sara Thomas. 2016. From First Footing to Faeries: An Inventory of Scotland's Living Culture. In *Patrimoine culturel immatériel et numérique,* ed. Marta Severo and Séverine Cachat, 199–206. Paris: L'Harmattan.

Otlet, Paul. 1934. *Traité de documentation: le livre sur le livre, théorie et pratique.* Bruxelles: Editiones Mundaneum.

Pietrobruno, Sheenagh. 2013. YouTube and the Social Archiving of Intangible Heritage. *New Media & Society* 15 (8): 1259–1276.

Severo, Marta, and Alberto Romele. 2015. *Traces numériques et territoires.* Paris: Presses de Mines.

UNESCO. 2003. Convention pour la sauvegarde du patrimoine culturel immaterial 2003. Accessed October 1, 2017. http://portal.unesco.org/en/ev.php-URL_ID=17716&URL_DO=DO_TOPIC&URL_SECTION=201.html

Wittgenstein, Ludwig. 1986. *Philosophical Investigations.* Oxford: Blackwell Publishing.

Part III

Digital Media Beyond Recording

The Unbearable Lightness (and Heaviness) of Being Digital

Stacey O'Neal Irwin

"Visible and mobile, my body is a thing among things; it's caught in the fabric of the world, and its cohesion is that of a thing. But, because it moves itself and sees, it holds things in a circle around itself" Merleau-Ponty (1974, 274).

1   Introduction

Technology is not neutral. Every use brings with it affect. As Maurice Merleau-Ponty's orienting quotation suggests, the human body is caught in the fabric of a world that enmeshes us. In contemporary society, this fabric is knit with a wide array of digital technologies presenting themselves in a variety of known and unknown ways. But the experience is not neutral. I describe this fabric of entanglement as the "digital attitude." This reflection starts with the human, the I, the digital user, who is part of the world in a priori, knowing things without having to think about

S. O. Irwin (✉)
Millersville University, Lancaster, PA, USA
e-mail: stacey.irwin@millersville.edu

A. Romele, E. Terrone (eds.), *Towards a Philosophy of Digital Media*,
https://doi.org/10.1007/978-3-319-75759-9_10

185

them. The aim of this reflective chapter is twofold: to push against the taken for granted attitude in the lifeworld to consider the human–technology–world experience, and to highlight some new ways of thinking about the digital environment. Interpreting digital media through the ideas of lightness and heaviness will be a central component of the chapter. This reflection is also paired with ideas from phenomenology and postphenomenology, to reveal themes of the digital attitude.

Experiencing the world is a process of knowing. This knowing, called epistemology, means understanding and articulating what we know and how we know it and why we know it. A portion of this worldly experience is taken for granted because it is so "known" that we do not need to reconsider or rethink our stance. And the taken for granted in the experience is often deeply seeded. In this coupling between a priori and taken for granted, sits the digital attitude, an intertwining of body in technology. This is where my exploration begins.

When digital technology is familiar, the body is often engaged in digital experience in a taken-for-granted way. Jaron Lanier, in his book *You Are Not a Gadget: A Manifesto* (2010, 187), explains:

> There is something extraordinary that you might care to notice when you are in VR [virtual reality], though nothing compels you to: you are no longer aware of your physical body. Your brain has accepted the avatar as your body. The only difference between your body and the rest of the reality you are experiencing is that you already know how to control your body, so it happens automatically and subconsciously.

In an immersive digital-media technology like virtual reality the body is meant to be taken for granted in the perceptual experience. The technology is designed that way, and research and development has studied a multitude of means to help the user take the technological side of the experience for granted. But that does not mean the experience has not left an imprint on the user.

Alfred Schutz and Thomas Luckman, in their book *The Structures of the Life-world* (1973), explain the taken for granted as a given in the lifeworld, the place where universal givens are part of everyday experience. In a world with many variables, the taken-for-granted is something we can count on as familiar because we have integrated life experiences, along with its successes and failures, into the vernacular of the everyday. The taken for granted is a habit of living that becomes "constituted in interpretations of experience (that is to say, explications of the horizon)" (Schutz and Luckman 1973, 9–10). Once we realize something has been encompassed into our everyday life as a taken-for-granted habit, we can examine it again. They explain, "by this taken-for-grantedness, we designate everything which we experience as unquestionable; every state of affairs is for us unproblematic until further notice" (Schutz and Luckman 1973, 3–4). Shultz and Luckman further share that humans understand their lifeworld to the degree necessary in order to be able to act in it and operate upon it. Day-to-day activities in the contemporary world contain an increasing abundance of technology. And the natural attitude has continued to encompass digital activity in known and unknown, unperceived ways to create the digital attitude.

And while digital technology may be seen as a cultural or social object within the lifeworld, it has increasingly become part of the fabric of the world, and less of an examined experience within it. Edmund Husserl's idea of *Lebenswelt* describes the lifeworld mentioned earlier as a world of everyday practices and truths. This is a world of straightforward experience between humans and all of the subjects and objects in that world as lived through the many perceptions of the body (Husserl 1970). Heidegger further studies *Lebenswelt* and makes distinctions between the physical and scientific world (*Umwelt*) and the social world (*Mitwelt*). These two distinctions further identify lifeworld experiences as objective or as phenomenologically lived. In addition, Schultz's explanation of *Lebenswelt* differs from Heidegger's version. Explains Kurt Wolff, in his text *Alfred Schutz: Appraisals and Developments;*

The *Lebenswelt* is said to be the linguistic structure that is presupposed for scientific, aesthetic and social discourse; practically it is that through which society, culture and personality are mediated. For Schutz, the social world was a world of shared meanings (Wolff 2012, 130).

Shultz's *Lebenswelt* relies on mediation within the world experience, which further connects with the digital attitude and the digital media habit. But what seems more distinct for this chapter is Schutz's different distinctions about *Umwelt* and *Mitwelt*. For him, *Umwelt* describes face-to-face daily interactions with those we are in close physical proximity with; and *Mitwelt* encompasses our interactions through a mediated connection in the lifeworld. Either way, we are humans, interacting in our world. But the distinction of the *Mitwelt* brings up ideas of collaboration and seclusion, individualism and distinctiveness versus conformity and collective thinking. For Schutz, the lifeworld always encompasses face-to-face and mediated interactions. In the digital world, these ideas intertwine in the fabric of the lifeworld. Examples include "face-to-face" video chat interfaces that bridge physical space, giving hearts or likes to posts on a "social site," collaborating on a big project while sitting alone in a secluded coffee shop working with people you will never see face-to-face, or collectively sharing opinions in a "hive mind" on a wiki or sharing website. Each example situates the body and the technology differently, but all are the *Mitwelt* kind of lifeworld. Using digital media is an existential dichotomy. It brings so much together but can also leave so much out of an experience. Embodied experience might be backgrounded in one digital media experience and very central to another experience. Humans can be physically alone while being social online, or existentially and technologically transported to another place while sitting at home. From robots to interfaces to augmented reality, the research and design world is working to bridge the gap between human and technology to create a more immersive experience. Digital media, the mediated connection, becomes a techno-social and techno-cultural experience for users.

Even more important is to question some aspects of the unquestioned, because the "structures of the lifeworld become apprehended as the fabric of meaning taken for granted in the natural attitude, the basic context of 'what is unquestioned'—and in this sense what is "taken as self evident'—that undergirds all social life and action" (Schutz and Luckman 1973, xxviii). The focus is to question the self-evident and unquestioned to begin to understand "digital" differently.

Digital media were not always a taken-for-granted experience, and still there are many experiences that involve intense thought and knowledge when working with digital technology. But as the digital divide closes, the digital natives grow up, and the technological becomes increasingly part of the lived world, the experience become increasingly taken for granted. Digital media designed to gather data on everyday life experiences mean that while some experiences are foregrounded, others are backgrounded at the very same moment. This compound experience creates a new dynamic to our technological understanding and to the human-technology-world experience. The digital becomes intertwined in the fabric of the lifeworld, the mundane world of the lived and unquestioned experience. This is where the digital attitude "lives."

2   Defining the Digital

Digital media is the "the contemporary name for digital content and digital devices like smart phones, tablets, computers, televisions, players, watches, gaming consoles and even billboards" (Irwin 2016, 17). Connection is key to digital media because the content plus the device creates a user-centered focus through an easy-to-use device to spread information through the Internet. The use and spread of digital media content and associated devices have embedded into the common-sense experience of the lifeworld for many. They are part of our digital attitude.

It becomes difficult to study digital media because, as a researcher, I also use digital media and may habitually take the experience for granted. As an example, I wake up in the morning by the alarm on my cell phone. I pop on the music app and listen to my preferred playlist as I prepare for my day. Somewhere in there I turn to my personal and work e-mail accounts to check on the digital correspondences that may have happened in the hours that I have slept. In the midst of this I receive text messages and phone calls. I look at my device's calendar and see I have reminders of meetings; my device alerts me that money was taken from my bank account to pay for the service of my device for the next month. I check the weather app and the news messages scrolling down the screen of my phone from my personally selected news sources. And I never once stop to consider, "Wow! I have a digital attitude." This is just my lifeworld, co-shaped by digital media. I incorporate digital media into my own body and my life experience in a habitual way. Merleau-Ponty explains the experience: "To get used to a hat, a car or a stick is to be transplanted into them, or conversely, to incorporate them into the bulk of our own body. Habit expressed our power of dilating our being-in-the-world" (Merleau-Ponty 2000, 143). Digital media have become the proverbial hat or coat or stick in Merleau-Ponty's example. Habit creates a "taken-for-grantedness" that covers over and alters the "transplanted" experience.

When we try to analyze it, we transpose these objects into consciousness. We commit what psychologists call 'the experience error,' which means that what we know to be in things themselves we immediately take as being in our consciousness of them. We make perceptions out of things perceived. We are caught up in the world and we do not succeed in extricating ourselves from it in order to achieve consciousness of the world (Merleau-Ponty 2000, 5).

This is what it is like to be in the world with digital media. To really dig into a use case like digital media, it is important to find a theoretical underpinning and methodology that can work toward revealing our relationship to and with technology. I use the underpinning scheme of phenomenology, a philosophical idea that studies the taken-for-granted experiences in the world, to further consider the digital attitude.

Phenomenology, in an initial and oversimple sense, may be characterized as a philosophical style that emphasizes a certain interpretation of human experience and that, in particular, concerns perception and bodily activity (Ihde 1990, 23).

To be good phenomenologists, we need to come upon a question we wish to study, and to "discover the origin of the object at the very center of our experience; we must describe the emergence of being and we must understand how, paradoxically, there is for us an in-itself" (Merleau-Ponty 2000, 71). We need to dig deep to find the "in itself" experience of digital media. Next we study some of the practices, experiences and interpretations of digital media through postphenomenology. The digital signal, in its most basic form, is a series of zeros and ones that repeat to form a variety of information to create digital media. "So dominant is this embeddedness of human-technology interfacing that from waking (alarm clocks and so on) to toilet activities (whole systems of water and sewerage) to eating (microwaves). There is technological involvement. This texture becomes the dominant, familiar and taken-for-granted activity of this 'world' of human inhabitants" (Ihde 1983, 73).

Again, the taken-for-granted is central in the human-technology experience. Digital media are not neutral. They have effect. And while it is evident that technology is not neutral, the enmeshed experience proliferates in a variety of ways. The digital component in digital media allows for exponential creation, sharing, and connecting and recreating through devices and platforms in many different ways. The media are forms of communication and expression but also an economic industry. When combined, they are a medium, something between other things like humans and technology. This creates a unique human–technology–world experience. "What stands out firstly is that all human-technology relations are two-way relations. Insofar as I use or employ a technology, I am used by and employed by that technology as well. There is a symmetry between humans and non-humans. A scientific instrument that did not or could not translate what it comes in contact with back into humanly understandable or perceivable range would be worthless" (Ihde 2002, 137–8).

Understanding the objective and subjective knowledge we have of digital media, the experiences we have with it, and the ways we experience it, helps us to better understand the human–technology–world experience. In fact, digital media can act as a model for understanding technological mediation. Specifically, the wide use and the unique embodied experience of so many forms of digital media content and such variety of devices and platforms and global connectivity create a technologically mediated pluraculture (Ihde 1995). After exploring practices, experiences and interpretations of digital media through phenomenology, a move to postphenomenology can study the enviornment of our technologically mediated pluraculture.

3   Postphenomenological Framework

Ideas used in the postphenomenological framework illustrate the co-shaped environment and the human-technology experience within it. "Postphenomenology, as developed first and foremost by Don Ihde, stays grounded in classic phenomenology, hermeneutics, and pragmatism, and so still revolves around central notions such as perception, embodiment, practice (or praxis), experience, and interpretation. But it 'updates' these notions in order to put them to work for the philosophical study of technologies and their usage" (Van Den Eede 2017, xviii). This kind of philosophy of technology, also called mediation theory, is rooted in the understanding of the world through a model called the I-technology-world schema (Rosenberger and Verbeek 2015; Verbeek 2016). The aim is to understand how technology shapes relationships between human beings and their world. The schema moves through four different relational shifts on a continuum that defines how the human and the technology connect through and in the technology. In its most basic relation, the experience can be seen as I-technology-world. Philosopher Don Ihde (1990) suggests that there are focal relations, embodied relations, hermeneutic relations, and alterity relations. Each of these models reveals something a bit different in the relation between the human (I) and the technology.

The idea of focal relations leads to the notion of technology in the foreground for use in a specific task, as distinct from background technologies that we do not think much about when they are working. In contemporary society we might consider such things as electricity, Internet access, and heating or air conditioning as background technology. These technologies have subtle but direct effects. When we are taking a quiz on a social-media app, much is going on in the background. While we answer questions, data points are connecting with an algorithm that creates information about the users' habits and interests and potential buying power. The data, once connected to prior use, can predict the time of day the user generally uses social media, and the additional places the user goes after a social-media site. A direct click from a social-media site to an outside digital location may also be recorded. All of these data points become something much larger and significant when they are composited to create a user profile. But they are in the background during use.

The (I-technology) → world schema explains the embodiment experience. When embodied, the user is experiencing perception through the technology. For instance, when the technology mediates the experience of the world, the user has incorporated the technology into action. Embodied technologies include glasses, a telescope, or a hearing aid. These technologies withdraw perception and the user experiences the embodied world through the technology. The (I-technology) → world schema is experienced through virtual-reality goggles. The user is "transported" through experience with the aid of the goggles. If the goggles are comfortable and well fitting, and the headphones are working, the augmented or virtual experience is an intertwining of humans and technology toward the world.

The I → (technology-world) model highlights when a human is interpreting or reading the world through the technology. This illustrates hermeneutic relations. Telling time on a clock or navigating a ship through instrument readings creates a hermeneutic reading of the technology. Digital-media examples include Global Positioning System programs that use graphics and satellite data points to translate location and give appropriate directions, or a wearable watch that tracks location, activity, and other diagnostics such as calories burned or miles walked.

Next, the I → technology (-world) model demonstrates alterity relations, the quasi-other or Technological Other experiences felt when completely enmeshed with the technology (see, for instance, Irwin 2005). The technological artifact shows itself as something other. A digital artist or editor who uses a variety of film or digital-editing software will anthropomorphize the experience and relate to and with the technology as a collaborative other, in a combined effort to complete an artistic task. Another example is Artificial intelligence (AI). The alterity experience is likened to meeting an Other similar to oneself; for example, in the quasi-othering experience users have with many computer devices such as an in-car navigation system or building something in a video game.

These models help explain the relation between the human and the technology in very specific and unique ways. What is interesting is how so many of these schema can be layered upon each other for use at the same time. For instance, I rely on Wi-Fi or cell reception (background relations), while I check my watch (hermeneutic and embodied) to learn my heart rate. I realize I must speed up my steps to get across town so I "listen" to my watch and increase my walking rate. I do not even remember that my watch is on all day, quietly and constantly quantifying my (alterity) movement when I am not even thinking about it. And I will look at my watch and think that it may not be working (alterity) if it does not seem as though I complete enough activity for the distance. I may blame my lack of quantified data on the watch itself, as if the fault were its own. At night, I check my watch to learn how my day went. My life-world has become *Mitwelt*, a world socially and culturally mediated by technology. Many different human-technology experiences overlap as they co-shape experience with technology.

Postphenomenology's variational theory is always a part of an analysis: "Variational analyses can show multistabilities, those distinctions and differences that may or may not be obvious at first glance, but also can be an analysis framework in itself" (Irwin 2016, 39). Don Ihde breaks variational distinctions into five different categories, including "the materiality of the technology, bodily techniques of use, cultural content of the practice, embodiment in trained practice and the appearance of differently structured lifeworlds relative to historical cultures and environments (2009, 19)." Once specific distinctions are considered, identifying a pivot point further names a specific practice or technology that highlights a viewpoint, position or standpoint to study. A pivot helps to identify multiple variations from one stable point. A pivot "stresses the degree to which the material of the artifact and human attentions can create different uses" (Whyte 2015, 76). The stability reveals the variations, which may include a particular use or part of the technology or experience of using the technology. These are called multistabilities, which explain that the very structure of technologies is multistable with respect to uses, cultural embeddedness, politics and ethics, among other things. But there is more to it. Multistability can be studied more specifically through micro and macro-perceptions. Micro-perceptions focus on body position, highlighting the sensory and instinctive experiences. Macro-perceptions provide perspective and environment for micro-perception, while highlighting multiple socio-cultural use contexts. After reflecting on phenomenological and postphenomenological considerations, exploring the lightness and heaviness of digital media complete the analysis.

4   Digital-for-Granted

Sometimes the study of opposites, in light of a specific technology, can reveal interesting patterns of use and experience. For instance, studying the lightness and heaviness of digital media can provide different perspectives on the technological use and experience. Lightness is defined as having little weight, a decreased significance, or to shine in brightness or radiance. Heaviness, in contrast, means having great weight, something difficult or severe, something depressing, or someone who is burdened or indifferent. One exercise to better understand the digital attitude is to play on some specific aspect or aspects of it. This reflection ends with a consideration of the lightness and heaviness of digital media. Illuminations, brightness, natural light, lightness of human awareness, truthfulness, and the lightness and heaviness of play are all studied by phenomenologists.

Heidegger (1993), in his exploration of the lifeworld, studies play as ease and lightness, and freedom from burden. Gadamer (2000) wrote of letting the thing stand in the light in his focus on play. His idea of play explores the seriousness and heaviness of practice and transformation. The ease in recording digital material and information, the ease in manipulating digital content, the difficulty of identifying ownership, and the ease in losing or erasing content are often taken for granted in the use. Both lightness and its converse, heaviness, can become tools in the study of the "in itself" experience of digital media.

Heidegger (1993, 172) explains, "a stone presses down and manifests its heaviness. But while this heaviness exerts an opposing pressure upon us it denies us any penetration into it [. . .] If we try to lay hold of the stone's heaviness in another way, by placing the stone on a balance, we merely bring the heaviness into the form of a calculated weight." His focus is to try to think differently about experience. This is my aim, too. The next section delves more deeply into the ideas of lightness and heaviness in the digital media environment. This list is not exhaustive but gives some concrete examples of the various trajectories of experiencing digital media.

4.1   Digital Lightness

Digital lightness means small, light, traceless, easy and illuminated. Digital lightness can be erased or lost more easily. Some digital trails are less easily traced and they evade the standard collection and surveillance devices. They carry a lighter footprint and can be easily transferred in undetected ways. Light digital files are less of a commitment, but also, if not backed up, are lost easily. Several countries are battling large search engines for the "right to be forgotten." The wish to tread lightly, in fact, non-existently, through the digital world is a human-rights issue for some countries. Those wishing for digital items written about them or documenting some fact about them to be removed from the digital environment's tendrils are facing an increasingly difficult task. Search engines thrive on linking data together for as long as the information, image or other content is available. While some traces become historically less available and can be lost before they were ever known, others are purposefully erased and are expected to stay that way as a kind of digital expungement. These issues and more become the everyday "fodder" of endless pieces of content generated and created and deleted and re-uploaded with increasing speed. Digital content that is taken less seriously is more taken-for-granted. This content becomes ownerless, easily reproducible and more globally transferable. The idea of music on a social sound platform like SoundCloud comes to mind. Non-copyrighted digital music is uploaded and downloaded and re-uploaded and altered and sampled to the point where ownership is no longer detected, though people do brand their version of hypergenerated and highly mixed un-owned freely available content.

We can look at a variety of themes that suggest the lightness of contemporary digital technology and the content it generates. They include the technological materiality of the device, noise, attention, embodiment, surveillance, toys and the multiplicity of use.

The direction of the nature of the materiality of digital technology is an evolution toward increasingly smaller devices. Music players and recording devices have shifted from inhabiting a corner of one's living area to fitting in the palm of one's hand. And in the case of some smart phones and digital audio players, even smaller than one's palm. These devices require smaller amounts of packaging and lighter-weight packages, which increase the lightness of resource use.

Noise issues can highlight digital lightness because decreased noise occurs through the use of headphones and earbuds. Gone are the days of blaring boom boxes carried around town. Decreased noise pollution creates less environmental noise, which promotes lighter (quieter) sounding environments and fewer competing noises in our environments. Noise-cancelling headphones completely eradicate unwanted environmental sound altogether.

Light devices decrease attention to themselves. Smaller devices tend to move to the background of our attentive area. All devices, when they become smaller, become less noticeable and blend into the background more readily. They fit on (and in) the body in a variety of ways for varied use. From small watches and implants to the Internet of Things, tiny is in. And less is more. Most digital media users carry a device that encompasses a multiplicity of uses. In contemporary society many people listen to music and watch media content by device without thinking about it.

The experience is not much more taxing than carrying around a wallet or deck of playing cards.

The lightness of the digital occurs because digital media devices multi-task many things with one device. Televisions have become multitools streaming music and entertainment content, so there is no need to have a DVD player or radio. Large stereo systems have been replaced with a mobile phone and a Bluetooth speaker that delivers better sound than a stereo console ever did. Yet for some, this lightness carries with it nostalgia for vinyl and record players, which evoke a psychological heaviness. Another example is the smartphone, which delivers the ability to monitor healthy habits, finances and banking, children and refrigerator contents while being social, playing games, listening to music, watching TV, and texting, talking and taking photos. In addition, having many devices is often more expensive and less convenient, than having one device with many uses.

Digital devices are so ubiquitous and hard to detect that camera and digital tracking, digital recording, and other kinds of drone-like surveillance can be virtually incognito. The lightness of digital toys makes them a great distraction and entertainment opportunity for children of all ages. Once engaged with games or media entertainment, children become quiet and digitally focused for at least a bit, until batteries go dead. These toys can fit in small places and provide varied entertainment for children for hours, then be recharged and hidden away for future use.

The lightness (and smallness) of digital technologies seems to de-emphasize their impact on culture and society. How can such a small piece of plastic and silicon Internet and communication technology (ICT) make socio-cultural changes in the world?

Some digital-media technologies are almost invisible. They are so small that they can be used without people even detecting their use, and sometimes people believe that others cannot see them use their devices in places where rules dictate that digital-media devices be stowed or put away. Users often do not understand that when they are texting on the sly, or using digital devices when they are not supposed to, that they are detected anyway, through the digital patterns they are emitting or, more often, the body language they display. In addition, recording others actions and behaviors as "proof" of completion has become increasingly popular. The digital footprint and transfer of the content is also so light that it is not detected when it is transferred through Wi-Fi or shared "live."

One further kind of digital-media lightness involves the ease of connectivity or use. Many devices "talk to each other" in a light and easy way which increases ease of use. No heavy amount of time is needed to master use. And instead of pushing buttons a user "touches" a screen. This light "touch" induces a different kind of use than button and lever pushing.

4.2   Digital Heaviness

In contrast, the theme of heaviness brings forward ways of thinking about digital media through use context, surveillance and background technologies like electricity, cultural shift, cost and perpetual use. As a quick introduction to digital heaviness, we can consider the proliferation of manufacturing and use. Digital content and materials mean more storage needs. The digital trace, and heat generated from this technology, increasingly connects to registration, collection and/or monitoring. The digital trace becomes a commodity in itself, creating new markets for the easy-to-transfer digital footprint. And who owns the digital footprint? Do I own my own content and digital emissions? Can someone use them? The heaviness continues when considering legal and ethical issues for digital content like music. Then, the digital environment needs to be "policed" in new and different ways, which increases the politics of the digital. Cutting, pasting, riffing, sampling. What is mine and what is yours? Do I have a digital marker, a fingerprint so to speak on my own digital stuff? Experts can trace ownership to a point but locating the digital signatures requires digital expertise, which becomes a new area of knowledge. Important and highly personal content becomes a commodity to be secured and stored for later access. And the music industry itself has shifted, as brick and mortar music stores fade and consumers move to downloading single songs, and then to buying streaming services that allow for personalized playlists. The technological movement resituates who owns what and who shares what.

Methods of collecting data have become increasingly easy to use and triangulate. Digital-media devices connected to data-mining software increase content collection through heavy device use. The more devices are used, the more information can be collected from that use. Gone are the days when a user can elect not to store website "cookies" in hopes of slowing down the collection of one's own personal data. Just clicking on a website creates content to save and analyze. User and consumer demographics and analytics are more available and more accurate than ever before. Additionally, heavy use is proliferated by social-media habits or constantly checking e-mails, texts, updates, and many other kinds of user generated content. While multitasking, many uses for one device increases convenience and ease; losing a device and trying to complete all of those tasks without it becomes a very heavy endeavor. The lack of back-ups, clouds, and other saving mechanisms makes some digital media experiences heavy with headaches. Multiplicity of uses adds cognitive clutter and at the same time clears physical space.

Surveillance carries with it both light and heaviness themes. There is an emotional heaviness to knowing the content or images that are recorded by certain individuals. What should be done with content that is private, and also possibly illegal, once it is covertly recorded?

A digital environment requires, at least for now, the heavy use of electricity. Digital devices entail a constant and heavy reliance of this utility, as do Wi-Fi and cloud servers. An emergency or natural disaster that cuts off power can quickly take away communication and create a heavy burden on those trying to receive help. Other forms of communication that are less reliant on power are no longer readily available in much of the world.

When face-to-face communication is largely replaced by deviced communication, important social rituals are altered. Eating, socializing, working, and playing have shifted to experiences mediated through technology. Bodies look down at devices instead of up at faces and eliminate other relational experiences like smiles (except by emoji), body language, and avoiding safety hazards.

The Internet shift from decentralized information to a recentralized framework alters access to information and decreases fringe movements and small community hubs. Large US technology companies like Google, Amazon, Facebook and Apple weld a heft of economic power largely due to the merging of smaller companies into larger ones and the data these companies have access to and collect. These conglomerates, and others similar to them, are being scrutinized for anti-competitive practices, tax evasion and invasion of privacy. Courts across the world are just beginning to scratch the surface of these companies' potentially problematic practices. This is digital heaviness.

Technology costs money and devices live and die by the upgrade. Devices become routinely outdated to keep the economy running, and reputations are made and lost on "who" has the newest "what." This heavy financial burden is a high payment for societal costs. And last, perpetual use of digital media creates a steady stream of material that measures and defines activity to the point where devices themselves, based on algorithmic data points, advise users to increase activity, drink more water, or increase heart rate. For some users, a doctor can monitor the experience from an online hub. This connectivity further increases the heaviness in reliability for health use, as blood cells, sugar intake, and beats per minute are counted and analyzed.

5   Conclusion: Digital 5.0

This reflection had two intentions: to explore the taken-for-granted experience of using digital media and to highlight some new ways of thinking about the digitally connected world. The aim of this kind of reflection is not to study or judge digital media as good/bad or positive/negative but to reveal non-neutrality in use and experience.

In the early days of Internet development, researchers and consumers alike wondered what Web 2.0 would be like. And then what Web 3.0 would encompass. And so on. What might something like Digital 5.0 look like? And how will the human–technology–world experience manifest? Postphenomenologists are currently studying the in-between of technological mediation (for instance, Van Den Eede 2011). But the opportunity to create empirical understanding of digital media earlier in the process of digital development could be even more fruitful if it occurred in the research-and-development (R&D) stage and not in the reactionary effects-driven phase.

As Don Ihde suggested, these kinds of philosophical renderings can be helpful in R&D. Philosophers too often undertake their reflections after the technologies are in place. Rather, they should reposition themselves at what Ihde calls the "R&D position" where technologies are taking developmental shape, in think tanks, in incubator facilities, in research centers. Only then can truly "new" and emerging technologies be fully philosophically engaged. In many cases philosophers are chasing ideas that have already been technologically solidified in the I-technology-world experience. Notes Ihde: "For an ethicist to try to determine what is the best allocation and fairest distribution of systems already in place or of effects already established, is in effect, to play a 'triage or ambulance corps' job after the battlefield is already strewn with the wounded and dying" (2003, 7).

This rings true for digital media as well. Situating philosophers in the beginning stages of developing digital experiences, environments and technologies or at least the next stages, can help to contribute to what may become Digital 5.0. Playing out theoretical use cases for technology can identify the multiple variations to consider unintended uses and consequences. An emphasis on the social roles of science and technology could further shape an understanding of politics and culture. Studying humans and the technology in relation, and how humans and the world are co-shaped through the technologies, can highlight a variety of human experiences and practices before they are solidified in design and usage. Specifically, taking actual technologies and technological approaches as a starting point for philosophical analysis, and moving to combine philosophical analysis with empirical investigation, yields a "philosophy *from* technology" or a "philosophy *for* technology" rather than a "philosophy *of* technology" (Verbeek 2016, 2). But we are not there yet.

An understanding of technological mediation through digital media leads to a better perception of the digital attitude, and the non-neutrality of our digitally mediated world. Themes of lightness and heaviness illustrate the non-neutral experience of the multilayered co-shaping digital environment, and contribute to further understanding of the human–technology–world relations. The theme of lightness focused on technological materiality of the device, noise, attention, embodiment, surveillance, toys and the multiplicity of use. The theme of heaviness explored ways of thinking about digital media through use context, surveillance, and background technologies like electricity, cultural shift, cost and perpetual use. Together, in the midst of the light and heavy, in a taken for granted lifeworld, we create the digital attitude.

References

Gadamer, Hans Georg. 2000. *Truth and Method.* New York: Continuum.

Heidegger, Martin. 1993. *Basic Writings.* New York: Harper.

Husserl, Edmund. 1970. *The Crisis of European Sciences and Transcendental Phenomenology.* Evanston, IL: Northwestern University Press.

Ihde, Don. 1983. *Existential Technics.* Albany, NY: SUNY Press.

———. 1990. *Technology and the Lifeworld: From Garden to Earth.* Bloomington and Indianapolis: Indiana University Press.

———. 1995. *Postphenomenology: Essays in Postmodern Context.* Evanstone: Northwest University Press.

———. 2002. *Bodies in Technology.* Minneapolis: University of Minnesota Press.

———. 2003. Postphenomenology-Again? Aarhaus University. http://sts.au.dk/fileadmin/sts/publications/working_papers/Ihde_-_Postphenomenology_Again.pdf. Printed at Trøjborgtrykkeriet, The Faculty of Arts, University of Aarhus.

———. 2009. *Postphenomenology and Technoscience: The Peking University Lectures.* Albany, NY: SUNY Press.

Irwin, Stacey O'Neal. 2005. Technological Other/Quasi Other: Reflection on Lived Experience. *Human Studies* 28 (4): 453–467.

———. 2016. *Digital Media. Human-Technology Connection.* Lanham: Lexington Books.

Lanier, Jaron. 2010. *You are Not a Gadget: A Manifesto.* New York: Vintage Books.

Merleau-Ponty, Maurice. 1974. *"Eye and Mind." Phenomenology, Language and Sociology: Selected Essays of Maurice Merleau-Ponty.* London: Heinemann Educational.

———. 2000. *Phenomenology of Perception.* London: Routledge.

Rosenberger, Robert, and Peter-Paul Verbeek, eds. 2015. *Postphenomenological Investigations: Essays on Human-Technology Relations.* Lanham: Lexington Books.

Schutz, Alfred, and Thomas Luckman. 1973. *The Structures of the Life-World*. Vol. 1. Evanston, IL: Northwestern University Press.

Van Den Eede, Yoni. 2011. In Between Us: On the Transparency and Opacity of Technological Mediation. *Foundations of Science* 16 (2–3): 139–159.

———. 2017. The Mediumness of World: A Love Triangle of Postphenomenology, Media Ecology, and Object Oriented Philosophy. In *Postphenomenology and Media: Essays on Human-Media-World Relations*, ed. Yoni Van Den Eede, Stacey O'Neil Irwin, and Galit Wellner, 229–250. Lanham: Lexington Books.

Verbeek, Peter-Paul. 2016. Toward a Theory of Technological Mediation: A Program for Postphenomenological Research. In *Technoscience and Postphenomenology: The Manhattan Papers*, ed. Jan Kyrre Berg O. Friis and Robert P. Crease, 189–204. Lanham: Lexington Books.

Whyte, Kyle Powys. 2015. What is Multistability? A Theory of the Keystone Concept of Postphenomenological Research. In *Technoscience and Postphenomenology: The Manhattan Papers*, ed. Jan Kyrre Berg O. Friis and Robert Crease, 69–82. Lanham: Lexington Books.

Wolff, Kurt. 2012. *Alfred Schutz: Appraisals and Developments*. Dordrecht: Nijhoff.

From Cellphones to Machine Learning. A Shift in the Role of the User in Algorithmic Writing

Galit Wellner

1   Introduction

Cellphones are now everywhere. Pictures on the Internet show Buddhist monks talking to a device in exotic locations, Wall Street bankers in elegant offices texting, obese foodies in greasy restaurants taking pictures and nervous parents in playgrounds looking at the cellphone's screen while hugging a screaming toddler. This is a partial list. Numerous people from various cultures, more than half of the world's population, use cellphones. And while many of these pictures show people talking on the cellphone, today, in the second decade of the twenty-first century, the handset is less frequently held near the ear and is more likely to be held by the hand in front of the user's face. Instead of talking, more and more usages are based on writing. In my book *A Postphenomenological Inquiry of Cellphones* (2015), I identified this move as a shift from a voice paradigm to a writing paradigm.

G. Wellner (✉)
NB School of Design, Haifa, Israel

Tel Aviv University, Tel Aviv, Israel

© The Author(s) 2018
A. Romele, E. Terrone (eds.), *Towards a Philosophy of Digital Media*,
https://doi.org/10.1007/978-3-319-75759-9_11

I showed how the cellphone maintained its identity as a "cellphone" although it is less frequently used as a telephone and more like a "writing machine" (as well as an Internet access point, a music player and a navigator, to name but a few of its uses). The voice and writing paradigms differ not only in their modes (voice vs text) but also in their purposes. While voice is mainly used for communications with others (humans, or automatic answering machines and the like), text can be used not only for communications (for example, texting) but also for recording. Cellular writing has preserved these two modes: communicating and memorizing. Texting is a common practice of asynchronous communication; calendar entries, to-do lists and phone books are modes of memorization that are meant to be kept private and not necessarily communicated (see Wellner 2015, 35–48).

Writing for the sake of recording is not new. Archeological research has shown that the very early writing in Mesopotamia 6,000 years ago was intended to list, count and categorize (Goody 1977). It was intended to memorize a certain state of affairs rather than communicate ideas, events and so on. Writing from the start has been a tool for recording; that is, the preservation of memory. This is also the main claim in Plato's *Phaedrus*.

Lately, cellular writing has been changing. With the introduction of smart algorithms, writing is no longer uniquely performed by humans. Today ever increasing portions of the recordings are delegated to technologies, so that algorithms write texts describing weather forecasts, stock-market analyses, sports articles, to name just a few. It is a writing that does not care if the text is read or not. What is important is to write and to document as much as possible. It is an inscription for the sake of storage. Writing for writing, writing without reading.

In order to explore the changes in writing and media technologies I turn to postphenomenology, a branch of philosophy of technology that analyzes our relations with technologies and through them with the world. One of the basic postphenomenological tools is the scheme "I–technology–world." This scheme represents the understanding that we are most likely to experience the world around us through the mediation of technologies, from clothing and dwelling, through eyeglasses, cars and televisions to contemporary cellphones and computers. Postphenomenology serves here as my analytical framework to understand the transformations in the cellphone's technologies and more generally in our writing paradigm.

As writing is an action that is intertwined with media technologies, this chapter opens with a short genealogy of modern media technologies from the printing press to cellphones and algorithmic writing. The next section examines the various genealogical steps with the postphenomenological scheme and its various permutations. The last section provides an overview of algorithmic writing and attempts to assess the relations between human readers and these technologies.

2   A Short Genealogy of Modern Media Technologies

Media have been with humans since prehistory, as evidenced, for example, in cave drawings of animals and the documentation of the Moon cycles. But until the invention of the printing press, media had been predominantly handwritten by humans. The printing press was the first time in human history that a segment of the writing process was delegated to a machine. Therefore, my starting point for modern media technologies is the printing press. Modern media technologies are not limited to the printing press and include image-based media technologies such as photography. Photography not only records visual memories, but also assists in the mechanization of their distribution. The focus of this chapter is however on the stage of writing, and distribution will be touched upon only lightly.

The next evolutionary step is the "new media" as offered by the Internet and accompanied by computer-based technologies (see Manovich 2001). New media are defined as the convergence of communication networks, information technologies and content (Flew 2008). They are digital and characterized as manipulable, shared among many users, and impartial. Yoni Van Den Eede points to the affinity between new media and digital media and characterizes this media as objectifying and constructing (Van Den Eede 2012, 336–341). Digital media are objectifying because they enable the fast and reliable creation of a copy of oneself via a digital camera. Thus, one can see oneself, thereby becoming an object for a gaze. Digital media are also constructing as they enable each user to become an author and distribute the newly created media to the masses. Stacey Erwin prefers the term "digital media" over "new media" for describing the technologies involved rather than their newness, which is obviously temporary (Irwin 2016, 6). However, as "new media" became a widespread notion, I will use the two terms here interchangeably.

The next step can be spotted in the cellphone era hallmarked by the ubiquity of media, what I have termed "new new media" (Wellner 2011). The characteristics that differentiate them from "new media" are availability anytime anywhere, speed of creation and distribution, and integration of older media forms and formats. If a personal computer connected to the Internet is the emblem of new media, for "new new media" the platform is the cellphone.

These three steps can also be identified by their reference to reality: the first step is occupied with the documentation of reality as text (printing press), voice (gramophone and radio) or image (camera, television); the second step of new media gives rise to virtual reality, albeit its roots can be found in literature and cinema; and the third step wishes to augment reality by adding layers of information to it. Table 1 depicts the main differences between the three genealogical steps.

In this chapter I wish to develop a fourth step; let us name it "algorithmic media." Like new media they are shared by many users; like "new new media" they are ubiquitous and spread at a fast pace. In this step media are not necessarily produced by human subjects nor do they need to serve them. They are produced by and for algorithms, and tend to serve the goals of corporations and governments, indifferent to the wishes and targets of individuals. There is no point in talking about the death of the author; the author is an algorithm. Unlike new media and new new media which are delivered over a technological object (personal computer or a cellphone) with which an end user interacts, the technological artifact that produces algorithmic media is a server hidden from the users. Even the reference to

Table 1 From media to new media to new new media

	1	2	3
	Media	New media	"New new media"
Emblem technology	Printing press	PC	Cellphone
Relation to reality	Record the real	Virtual reality	Augmented reality

reality is different: in the algorithmic media space, reality is that which can be grasped by the algorithm. Such a reality exists in parallel to the reality as grasped by people, and it is an interesting question: What happens when these two realities meet or collapse into each other?

3   A Short Introduction to Postphenomenology

If phenomenology studies our experience in and of the world, postphenomenology studies how our experience of the world is mediated by technologies. The basic postphenomenological scheme describes the constitutive role of technologies as "I–technology–world." With the addition of arrows and parentheses, several permutations can be developed (Ihde 1990). Thus, when the "I" and the "technology" are within the parentheses, the scheme denotes embodiment relations in which the combination of me and the technology experiences the world as one unit. The arrow from the combination towards the "world" signifies an intentionality of the combination that is directed to the surrounding environment. The scheme looks like this:

$$(I - technology) \rightarrow world$$

This is the case of the pen and the keyboard that become a (temporary) part of our body schema during the writing process. When media are analyzed with embodiment relations, the writing process is examined from the viewpoint of how the writing tool is held on the bodily level. The analysis does not take into account the intellectual effort that is required at the writing phase, an effort that runs in parallel to the embodied effort. This is dealt by the hermeneutic relations that model the situation when the technology and the world function as one experiencing unit, and are consequently within the parentheses. The arrow from the "I" denotes that intentionality is directed towards that unit. In this process the "I" reads the technology in order to know the world and extract meaning, and hence the term "hermeneutics." The scheme for hermeneutic relations is:

$$I \rightarrow (\text{technology} - \text{world})$$

An example of hermeneutic relations in the context of media technologies can be watching news on television. The intentionality is directed to the combined unit of television–world and they are conceived together. The television is mediating the world and gives meaning and context to events which otherwise would not have been taken into account. A certain media literacy is required in order to properly "read" the role television plays in this process and identify its biases.

Embodiment and hermeneutics are the two main postphenomenological relations that deal with media and they will be discussed in details later in this section. To complete the picture I briefly mention two more relations: alterity and background relations. In alterity relations the world withdraws to the background and leaves the "I" to interact with the technology as a quasi-other (see Wellner 2014). Here the parenthesis function slightly differently and they denote a withdrawal of the world, thereby positioning it outside the focus of the relations. The scheme for alterity relations is:

$$I \rightarrow \text{technology} (-\text{world})$$

Our dialogue with an automated teller machine is an example of alterity relations. When we interact with this technology, we hardly pay attention to our surrounding. Lastly, in background relations it is the technology that withdraws to the background and becomes unnoticed, and is therefore marked in the scheme by the parenthesis:

$$I \rightarrow (\text{technology} -) \text{world}$$

Background relations usually happen when we are immersed in an activity like reading and do not pay attention to the technologies that function in the background. Table 2 depicts the four basic relations and their respective permutations of the postphenomenological scheme.

Table 2 The postphe-
nomenological
relations

Embodiment relations	(I–technology) → world
Hermeneutic relations	I → (technology–world)
Alterity relations	I → technology (–world)
Background relations	I → (technology–) world

In reading and writing, not only hermeneutic relations are relevant. Sometimes media also function as part of the experiencer's body, thereby maintaining embodiment relations in parallel to hermeneutic relations (see Ihde 2010, 128–39; Rosenberger 2008; Forss 2012). Embodiment relations highlight the technological element and disregard the content element(s). Therefore, hermeneutic relations are currently the main postphenomenological vehicle to understand our special association to media technologies. However, the existing postphenomenological models do not refer specifically to writing. They refer more to reading and interpretation and less to the "production" phase. More importantly, there is hardly any discussion on the outcomes of the act of writing which is, in practice, a change in the world.

This neglect can be explained by the very nature of hermeneutic relations, in which the world is seldom accessible directly; the world is mediated by a technology-text. Don Ihde explains: "the world is first transformed into a text, which in turn is read" (Ihde 1990, 92). Andrew Feenberg stretches Ihde's transformation ingredient and regards hermeneutic relations like a screenplay in which "the interpreted message stands in for the world, is in effect a world" (Feenberg 2006, 194). According to Ihde and Feenberg, "technology" and "world" are not just two parts of one logical unit but rather "technology" replaces "world" by a message-text (see also Wellner 2017). The role of the "I" is to impute meaning to an existing message-text.

As an interpretation-oriented relation, hermeneutic relations are centered on text. It is assumed that somebody wrote a text; that the text exists; and that now the "I" reads it. These are three stages stretching along a timeline: writing, displaying, and reading. Note that before the emergence of digital technologies, only writing and reading were relevant; digital technologies introduced a middle stage of display. Most of the postphenomenological analyses of hermeneutic relations focus on the latter stage of reading, and pay less attention to the earlier writing phase (see Wellner 2017). This focus on reading in hermeneutic relations follows the cultural inclination to "a literate worldview" which centers on the figure/content and leaves the ground/form in the background (Van Den Eede 2012, 167). The writing stage is left in the background partially modeled, with some elements covered by embodiment relations. In the next sections I will portray what writing relations may look like for each of the genealogical stages. I will demonstrate how the emergence of machine learning algorithms changes the writing relations and redefines the roles of the "I" and the "world."

4   Writing Relations across the Genealogical Steps

As I argued elsewhere, the postphenomenological hermeneutic relations are limited to reading and a new type of relation should be added to represent writing (Wellner 2017). In this section I describe each of the genealogical variations of modern media through the lens of writing relations. In order to do so, I look into some of the recent developments of the postphenomenological scheme and offer some of my own.

4.1   Printing Press

The first modern media is the printing press. This mode of writing heavily involves the human body for the arrangement of the letters, the placement of sheets of paper and the operation of the machine. But traditional writing with a pen or a pencil also involves the body (see Ihde (1990) and especially the examples on the invention of the alphabet). The differences between the two technologies and related technics can be demonstrated through Don Ihde's critique of Martin Heidegger's writing analysis.

In his Parmenides course, Heidegger positions handwriting as an authentic form of writing and thinking and hence the only one that is acceptable. He praises the human hand as a "tool" for thinking and asserts that "the word as script is handwriting" (Heidegger 1992, 80 [118–9]). Heidegger criticizes writing with a typewriter and points to this form of writing as "one of the major reasons for increasing destruction of the word" (Heidegger 1992, 81) because it takes out the personal element of writing. Typewriting can only be justified in cases where it is used for "preserv[ing] the writing" or to "turn into print something already written" (ibid). And he concludes: "mechanical writing deprives the hand of its rank in the realm of the written word and degrades the word to a means of communication" (ibid).

Heidegger's remark in the Parmenides course is based on his distinction between "traditional technologies" and "modern technologies" (Ihde 2010, 117), where the latter are blamed for "enframing." Ihde "turns Heidegger onto Heidegger" (Ihde 2010, 121) and criticizes Heidegger's hostility to the printed word. He offers us an imaginary tour to the time when Heidegger wrote his *Being and Time* with a pen:

> He is sitting at his desk, writing by hand [. . .] producing the text, *Being and Time*. He is rushing to complete it so that he can receive his promotion [. . .]. It works and the publisher accepts the manuscript for publication. But what then? Here we enter a context of ambiguity and eventually irony: before printing, the manuscript must be typed, reduced as it were, to the uniformity of print and standardization. Does this make Heidegger a human being just the same as every other human being? I think not [. . .]. The book, once read and evaluated [. . .] gets praised as one of the most important philosophical works of the twentieth century. In short, Heidegger could not have become Heidegger without the infrastructure of the Gutenberg Revolution of printing, publication and circulation. (Ihde 2010, 125)

Ihde's scenario clearly demonstrates that the printing press and the typewriter are essential technologies to modern writing processes, and the attempts to categorize them as constraints on thinking or authenticity are easily classified as romantic and dystopian.

In classical postphenomenology, writing is covered by a combination of hermeneutic and embodiment relations. The printing press as mode of writing heavily involves the human body as described above. Its examination may lead us to ask whose body is involved. Put differently, the printing press disjunctions between the writer and the end-result so that the printed manuscript is no longer "manu" but only a "script." The chain of writing process becomes longer and in-between the writer and the printed script there is also the operator of the machine who does not need to know how to read or write. This part of the mechanized writing process could be at its extreme form modeled by embodiment relations with no need for hermeneutic relations! The segmentation of the writing process should lead us to regard writing as separated from reading. Consequently, a new type of relation seems to be needed even in occasions where hermeneutic relations were considered sufficient.

4.2   New Media

My next genealogical stage is new media. Here the text is digital and the production processes are sometimes different from those described for the printing press. In the digital age, all stages require reading, which leads to an even greater analytical emphasis on hermeneutics. Postphenomenology, still focused on reading, found that digital reading is different from the analogue. One of the features of digital text that attracted postphenomenologists' attention was its active role and ability to influence (and modify) the reader's perception. Thus, Peter-Paul Verbeek (2008) offers some expansions to Ihde's four postphenomenological relations and develops a framework titled "cyborg intentionality." Of relevance here is his notion of "composite intentionality" that refers to situations in which intentionality is distributed between humans and technologies. The underlying idea is that technologies can have a certain form of intentionality. Don Ihde explains this as "instrumental intentionalities or built-in selectivities in technologies" (Ihde 2015, xv). And Robert Rosenberger and Peter-Paul Verbeek elaborate: "Intentionality is not a bridge between subject and object but a fountain from which the two of them emerge" (Rosenberger and Verbeek 2015, 12). This sense of intentionality is somewhat similar to ANT's supposition that non-human actors can have agency. Likewise, Verbeek describes a situation in which technological intentionality is not necessarily directed to accurately represent the world, but rather constructs a new way of seeing the world. He gives an example of an art work using a unique technique of photography to create an image of the world that is empty of anything that moves. The result is a world picture popoulated only by mute objects with no living creatures. In this example, the viewers and their act of reading are at the center. The artists and the techniques of producing that text-picture are a given. Verbeek provides another example of composite intentionality, this time the world representation is based on input from a stereographic camera and some manipulation on the photographs so that the result is (again) a fictitious world picture. But now it can be viewed only with the help of a 3D headset. He terms it "constructive intentionality" because the camera and the additional viewing technology construct a reality in such a way that the result is fictional. The artist has a god-like role of designing a certain world(view).

To represent the relations between these works of art, their viewers and the world as depicted in those works, Verbeek suggests replacing the dash sign between "technology" and "world" in the postphenomenological scheme with an arrow, thereby creating a permutation on the classical hermeneutic relations:

$$I \rightarrow (technology \rightarrow world)$$

Although not intended to model writing relations, Verbeek's permutation can also represent how writing technologies create a trace in the world. It can reflect the active role of a text as a trace that remains long after the act and sometimes even the writer.

Verbeek's examples refer to the artistic image and the equipment that enabled its creation and later its viewing. Both form together the "technology" element in the postphenomenological scheme. Heather Wiltse (2014) unpacks this element and suggests some adjustments to the hermeneutic relations in order to properly deal with digital texts (which she terms as digital materials). These adjustments mainly include a split of the "technology" element into substrate and trace, where substrate denotes the enabling medium on which the text is written (or otherwise inscribed) and trace indicates a content-text (or anything that can be interpreted). Wiltse explains that in digital environments the trace and the substrate are separated so that contents are produced by a device different from that with which they are displayed (although some traces are never displayed, like those generated by sensors that record move-ment). She demonstrates the trace-substrate dichotomy with the classi-cal example of hermeneutic relations: the thermometer. In its analogue version, the mercury measures and shows the temperature at the same time, simultaneously functioning as substrate and trace. Next, in a digi-tal thermometer, the measurement and the display are performed by two different components. Lastly, a weather website offers yet another break and fracture when it provides temperature information for remote locations, as if they were right outside the door. The technology that inscribes the information has no direct link to that which presents it. Wiltse concludes that the functional uncoupling of writing and dis-playing is paradigmatic to digital environments. This separation is rep-resented by a vertical separator in her variation to the postphenomenological scheme so that the substrate faces the "world" while the trace faces the "I":

$$I \rightarrow \left(\left[\text{trace}|\text{substrate} \right] \rightarrow \text{world} \right)$$

Her revised postphenomenological scheme covers all three stages of writing, displaying and reading. This scheme is intended to replace her-meneutic relations in the context of digital environments. But this revised formula does not reflect the separation between reading, writing and dis-playing as demonstrated by the thermometer technologies. Reading remains the dominant action. The purpose of this chapter is to show that although writing and reading in the computerized world are programmed separately and differently, these processes on the human part are still regarded as entangled and almost inseparable, as they were in the age of the printing press.

4.3   New New Media

The third genealogical step is based on the ubiquity of data as enabled by cellphones and other portable or wearable technologies. Here I find rel-evant the work of Nicola Liberati on augmented reality (2016), in which he differentiates between the physical technological artifact that produces the experience of augmented reality and non-physical technological representations which are the layers of information displayed on top of the reality view. In his terms, "technology" is the technological artifact and "object" or "label" is the technological representation, or the text. This divide reflects the Janus face of media as form and content, and resembles Wiltse's separation between trace and substrate. Liberati uses this distinction in order to examine attention during reading processes. His version of the postphenomenological scheme uses curly brackets to denote to where the user's attention is directed: that which is positioned within the curly brackets withdraws to the background. He presents a set of two permutations:

$$\text{Subject} \rightarrow \left(\{\text{Object}\} - \text{Technology} \right)$$

$$\text{Subject} \rightarrow \left(\text{Object} - \{\text{Technology}\} \right)$$

In the first instance the text withdraws to the background, and the user's attention is directed to the technological artifact: a paper, a screen or a writing tool. This situation may reflect a certain mode of writing that focuses on the format, as in Japanese calligraphy or the experience of those who begin to use a computer and look for the letters on the keyboard. In the second instance, the technological artifact that enables the reading/writing withdraws to the background. It is a variation of the hermeneutic relations in which the attention is focused on the text ("object" in Liberati's terminology) and can also represent a mode of writing in which the assisting technologies withdraw to the background and the user-writer concentrates on the text itself. Surprisingly, Liberati does not refer to the world component of the scheme and the changes it goes through in the writing process. This lack can be imputed to the nature of augmented reality technologies in which the borderline between the world and the text blurs, and both are experienced together or, better, read together. The focus on augmented-reality technologies dictates the primacy of reading. Writing in augmented reality is rarely performed by humans; it is more likely to be the product of algorithms, which are the subject of the next genealogical step.

4.4 Algorithmic Writing

Recently, machine learning algorithms have written texts such as weather-forecast articles, stock-market reports and sport-event coverage. These texts look as if a human journalist wrote them, and so the algorithms that produce them are frequently classified as artificial intelligence (AI). AI technologies call for a new conceptualization of the writing "I." In this genealogical step I analyze the relations between the humans involved in such writing processes and the AI algorithms that write.

Algorithmic writing is based on two sets of data. The first is used for training, and the second for the actual production serving as raw materials from which information is extracted and developed into a human-readable text. Needless to say, this text is not necessarily alphabetic text; it can be graphs and charts. Frequently, the data comes in huge amounts, beyond a human ability to read all of it, let alone carefully and attentively. The data set can be weather indications collected by satellites, car movements as depicted by cellphones and traffic sensors, or a stack of tens of thousands of emails released by Wikileaks (see Irwin 2016). The raw data is hardly read by a human being. One could expect the end result to be read, but here is the difference: the writing algorithm does not know and does not care if the end results are read by humans. It just writes. It is a writing that records a certain state of affairs. It is a recording for the sake of recording.

My attempt to model writing relations in the age of artificial intelligence requires a major change in the postphenomenological scheme. So far, all permutations have been based on human intentionality and so the arrow has gone from the "I" towards the "technology" and the "world." Now it is time to reverse the arrow and reflect the technological intentionality of the algorithms. Since algorithms differentiate between the software and the data, I will follow Wiltse and Liberati and split the "technology" element into "tech" (for technological artifact) and "text." Contra Wiltse and Liberati, I place "tech" first to be followed by the "text" so that "tech" is facing the "I," and not "text":

$$I \leftarrow \text{algorithm} \rightarrow \text{text} \rightarrow \text{world}$$

The algorithm is directed toward the writing of the text, and directs the reading "I" what to read. The third arrow located between "text" and "world" represents the creation of a trace in the world, as suggested by Verbeek. This permutation is non-anthropocentric and represents the increasingly important role of algorithms in our everydayness.

5   Writing Relations in the Digital Age

Machines have been writing texts for several decades. In his article "Perceiving other planets: Bodily experience, interpretation, and the Mars Orbiter camera" (2008), Robert Rosenberger discusses the relations between scientists-readers and machine writing as performed by the Mars Orbiter Camera (MOC) that produces pictures of the planet. The system modifies the pictures to incorporate data from other instruments such as thermal emission detector and laser altimeter. Rosenberger details a technology-intensive multi-phased translation process that runs from an analog real-world picture to a digital image, through its transmission in space, integration of information from various instruments into a single coherent image, and finally to a reverse translation back to a picture that can be seen by humans on a computer screen (rather than a long series of numbers). He shows how the combining algorithm shapes the content and allows multiple interpretations. Rosenberger invests much effort in explaining the technical details of the writing process of MOC's digital images, but his analysis is headed towards the reading stage and how humans interpreted the images. He remains committed to the centrality of reading and so the classical hermeneutic relations are his main analytical tool. In his article the writing phase is not modeled with the postphenomenological scheme. My suggested scheme can reveal the centrality of the MOC:

Scientists ← Mars Orbiter Camera → Pictures of Mars → Mars

Another relatively early form of machine writing can be found in the automatic writing of notifications, small pieces of text that can be displayed one-by-one or together as a tendency. Usually in the framework of mobile apps or social-networking platforms, notifications indicate an action, a location or a state-of-mind. Once the app writes a notification, the user (as well as others) can read the notification. The app CouchCashet (cited in Wiltse 2014) "checks-in" at a location of a party, even if the user did not attend it. The purpose is to build for the user a reputation as a "party animal," thereby turning the notification mechanism into a social signaling apparatus. The intentionality combines the user's and the technology's; they both participate in the writing process.

In many cases the automatic notification requires only the implicit consent of the user. When one logs in to a Wi-Fi network in an airport or a restaurant, for instance, some of these systems require or allow logging in via Facebook profile. Then the system posts a notification on that person's wall on Facebook that "person X visited place Y on date Z." Other apps decide when to post an "away" notice on an instant messaging app, thereby signaling to the world that the person is not available to talk or chat (Wiltse 2014, 173). It is possible that the person might still be at her desk, and it makes a difference who posted the notice—the algorithm or the user. If the algorithm posts the notice, that person might be able to receive a call or can be expected to answer a quick question; but if the person intentionally set her status to "away," it means she does not wish to be disturbed. In any case, the notification's text is relatively simple and does not require much creativity. The next types of writing come closer to what we term creativity and until recently classified as uniquely human.

Down the road of digital writing's evolution, machine writing composes a dynamic worldview that is updated in real time and changes all the time. Think of a video game or a map showing traffic conditions. These forms of digital writing match Lev Manovich's distinction between document and performance (2013), according to which a document is fixed and can be accessed in an identical way over and over again, while performance is the unstable representation we see when we access online navigation maps, scenes in games or real-time stock-exchange data. Because algorithms write these texts (frequently based on notifications as explained above), they keep changing according to the time they are produced and often according to the characteristics of the person who operates the app. Unlike traditional documents which one could read time and again, under the performance regime the displayed text cannot be re-read. It will always change the next time it is displayed.

If we agree with Manovich on the shift in media architectures, we need to understand how machines write. Let's return to journalism, where we can find robo-journalism in which algorithms scan huge amounts of data and quickly build articles of them. Today robo-journalism is done mostly in the areas of financial reporting and coverage of sport events. These areas were selected for their table-like data-structures so that algorithms can easily read them. It is interesting to note, however, that today's automated article-writing technics of machine learning have human origins. It means that human journalists train the system with sample writings and assess the end results, thereby "teaching" it how to write properly. Teaching how to write an article like a human is necessary as long as the articles are produced for human reading. While most human readers will not be able to notice if a financial news article was written by a human or by an algorithm, sometimes there is a difference. Ethical, moral and intellectual property issues arise. The suggested postphenomenological scheme for writing relations for robo-journalism looks like this:

Human reader $\leftarrow$ robo-journalist $\rightarrow$ article $\rightarrow$ world (including readers)

In this formula, human readers can be found in two elements, the "I" and the "world": as "I," the reader simply reads the article produced by the algorithm; as "world," the reader is part of a group that has a public opinion and has the potential of acting as a response to the article.

My analysis of robo-journalism purposefully does not refer to the developers of the algorithm. Machine learning algorithms are different from traditional algorithms that required the coding of if-then-else loops. In machine learning, the algorithm develops independently such loops, usually in a manner opaque to humans. The developers set the platform, and its direction is determined on the fly in the training phase with the help of examples. Hence these algorithms have a some autonomy and intentionality (see Just and Latzer [2017]). Autonomy and intentionality require a certain amount of cognition as a basis. Recently, cognition has been studied and modeled to allow attributing algorithms some form of cognition.

6   Conclusion

In this chapter I examined the writing paradigm of the cellphone and its successors. I mentioned the two purposes of writing—communication and memorization. The former is dominant in communications and media studies while the latter is elaborated mostly in philosophy of technology. My genealogy showed a movement between the two modes of writing: from the establishment of modern mass communication by the printing press, through computerized new media and augmented reality's "new new media," ending at algorithmic writing that records for the sake of recording, and only communicates as a by-product. This genealogy showed how we live in "algorithmic reality" that is different from "reality construct[ed] by traditional mass media" (Just and Latzer 2017, 238). The chapter offers preliminary insights to situations in which the two realities collide and in which writing is not reserved for humans.

The genealogy also served as a map for the developments in postphenomenology as a theory of technology through a review of various permutations to the classical hermeneutic relations. The analysis evolved from Ihde's conceptualization of hermeneutic relations, to Verbeek's composite intentionality, Wiltse's trace and substrate and Liberati's object and technology. My own contribution is in the reversal of the intentionality arrow, so it points to the human reflecting a strong technological intentionality.

I demonstrated the technological intentionality of smart algorithms with the examples of the MOC, check-in and automatic notifications and robo-journalism. These contemporary examples can be mapped at the higher end of today's algorithmic landscape. Michael Latzer builds a functional typology of algorithmic selection applications ranging from search algorithms, through aggregation, filtering, recommendation and so on, to content production, computational advertising, and algorithmic trading (Just and Latzer 2017). This chapter is focused on the content production. Indeed algorithms direct the web readers what to read so that conservatives would receive mainly texts supporting neo-liberalism. My aim in this chapter has been to go one level deeper and look at the production of the texts themselves. Paraphrasing Husserl, I would call this direction "back to the texts themselves." I would question the possibility of algorithmically writing of texts that will be creative, and that can lead the reader to think and rethink positions and beliefs.

References

Feenberg, Andrew. 2006. Active and Passive Bodies: Don Ihde's Phenomenology of the Body. In *Postphenomenology: A Critical Companion to Ihde*, ed. Evan Selinger, 189–196. Albany, NY: SUNY Press.

Flew, Terry. 2008. *New Media: An Introduction*. Oxford: Oxford University Press.

Forss, Anette. 2012 Cells and the (Imaginary) Patient: The Multistable Practitioner–Technology–Cell Interface in the Cytology Laboratory. *Medicine, Health Care and Philosophy* 15 (3, Aug.): 295–308.

Goody, Jack. 1977. *The Domestication of the Savage Mind*. Cambridge: Cambridge University Press.

Heidegger, Martin. 1992. *Parmenides*. Bloomington and Indianapolis: Inidiana University Press.

Ihde, Don. 1990. *Technology and the Lifeworld. From Garden to Earth*. Bloomington and Indianapolis: Indiana University Press.

———. 2010. *Heidegger's Technologies. Postphenomenological Perspectives*. New York: Fordham University Press.

———. 2015. Preface: Positioning Postphenomenology. In *Postphenomenological Investigations: Essays on Human-Technology Relations*, ed. Robert Rosenberger and Peter-Paul Verbeek, vii–xvi. Lanham: Lexington Books.

Irwin, Stacey O'Neal. 2016. *Digital Media. Human–Technology Connection*. Lanham: Lexington Books.

Just, Natascha, and Michael Latzer. 2017. Governance by Algorithms: Reality Construction by Algorithmic Selection on the Internet. *Media, Culture & Society* 39 (2): 238–258.

Liberati, Nicola. 2016. Augmented Reality and Ubiquitous Computing: the Hidden Potentialities of Augmented Reality. *AI & Society* 31 (1): 17–28.

Manovich, Lev. 2001. *The Language of New Media*. Cambridge and London: The MIT Press.

———. 2013. *Software Takes Command*. New York and London: Bloomsbery Academics.

Rosenberger, Robert. 2008. Perceiving Other Planets: Bodily Experience, Interpretation, and the Mars Orbiter Camera. *Human Studies* 31: 63–75.

Rosenberger, Robert, and Peter-Paul Verbeek. 2015. A Field Guide to Postphenomenology. In *Postphenomenological Investigations: Essays on Human-Technology Relations*, ed. Robert Rosenberger and Peter-Paul Verbeek, 9–41. Lanham: Lexington Books.

Van Den, Eede. 2012. *Yoni. Amor Technologiae. Marshall McLuhan as Philosopher of Technology—Toward a Philosophy of Human-media Relationships*. Brussels: VUBPress.

Verbeek, Peter-Paul. 2008. Cyborg intentionality. Rethinking the Phenomenology of Human–Technology Relations. *Phenomenology and Cognitive Science* 7: 387–395.

Wellner, Galit. 2011. Wall-Window-Screen: How the Cell Phone Mediates a Worldview for Us. *Humanities and Technology Review* 30: 77–104.

———. 2014. The Quasi-Face of the Cell Phone. Rethinking Alterity and Screens. *Human Studies* 37 (3): 299–316.

———. 2015. *A Postphenomenological Inquiry of Cell Phones. Genealogies, Meanings, and Becoming*. Lexington Books.

———. 2017. I-Media-World. The Algorithmic Shift from Hermeneutic Relations to Writing Relations. In *Postphenomenology and Media: Essays on Human–Media–World Relations*, ed. Yoni Van den Eede, Stacey Irwin, and Galit Wellner, 207–228. Lanham: Lexington Books.

Wiltse, Heather. 2014. Unpacking Digital Material Mediation. *Techné: Research in Philosophy and Technology* 18 (3): 154–182.

A Philosophy of "Doing" in the Digital

Stefano Gualeni

1 Introduction

US scholar Janet Murray observed that the way in which we customarily address digital media in their plural form is symptomatic of the ongoing academic confusion about the emergence of what is (in fact) a single medium (Murray 2003, 3). In this chapter, the terms "computer" and "digital medium" are used as synonyms, and will be used in their singular form.

A frequently invoked theoretical perspective on technologies of communication and representation proposes to understand all media as always "commenting, reproducing, and replacing" older media forms (Bolter and Grusin 2000, 55). Understood through the evolutionary perspective that was just outlined, the digital medium is recognized as offering, combining, and further advancing technical possibilities and ways of conveying information that were previously the exclusive domains of older technologies of communication, representation, and (playful)

S. Gualeni (✉)
Institute of Digital Games, University of Malta, Msida, Malta
e-mail: stefano.gualeni@um.edu.mt

interaction. To clarify this point, it might help to ground the multimediality of computers in rather dull examples taken from my own everyday existence. Just yesterday—on the battered laptop on which I am now typing these words—I called up an old friend in Chicago, looked up ways to alleviate the prickling of eczema, listened to a radio podcast, did some boring accounting, practiced a dozen Contract Bridge hands against an artificial intelligence in preparation for an upcoming tournament, and watched a Woody Allen movie.

If the obvious versatility of the digital medium did not make it hard enough already to produce comprehensive statements about its "character" (its experiential possibilities and effects), the philosophy of technology reminds us that (as for any other technologies) computers are also and always appropriated by their users in ways that are adaptable, context-dependent, and often unpredictable.[1]

To this already messy conceptual "soup" about what the digital medium is and what the digital medium does, I believe we need to add an "ingredient" that several scholarly approaches consider central to the "recipe." I am referring to the fact that computers disclose methods of fruition and production that favor the active participation of their users (Murray 1998; Calleja 2011). Resorting to a perhaps brutal generalization (or maybe a helpful lie) we could say that information and representations that are experienced through the digital medium are not simply received and interpreted, as was the case with pre-digital forms of mediation. The computer discloses contents in ways that are less passive and linear than pre-digital media, and empowers its users to explore, manipulate, and produce information (Manovich 2001; 2013).

This book promises to further our understanding of the digital medium by concentrating its attention on the recording and archiving capabilities of the computer. The verb "to record" finds its linguistic roots in the Latin *recordare*, composed of the particle *re-* (again, anew), and the verb *cordare* (from the word *cor*, "heart," which was considered to be the organ responsible for storing memories and eliciting passions). Etymologically, the verb "to record" refers to the act of re-visiting or re-membering something that was previously thought, learned, or experienced. Words, images, smells, feelings, ideas, and sounds that we re-corded in the past can be re-vived in our experiential and emotional present with a degree of fidelity that depends on the technical medium utilized.

Instead of largely focusing on understanding the computer as a medium for recording information, this chapter focuses on its qualities and possibilities as a technical mediator; that is to say, on its disclosing and inviting specific ways of acting, communicating, translating, experiencing, and understanding (Verbeek 2005). With this objective in mind, I deem it more fruitful to analyze the technical and experiential possibilities of the digital medium in ways that are oriented towards prefiguring and configuring the future, rather than archiving and interpreting the past.

Before that, I believe it is relevant to clarify that I do not wish to portray the "recording" qualities of computers as merely driven by desires to preserve our personal and collective history. In fact it is also (if not mainly) under the guidance of growing and increasingly more pervasive "digital records" of our past actions and choices that social policies, academic research, marketing campaigns, and industrial production decisions are pursued. By facilitating the gathering and interpretation of vast amounts of information, the computer allows manipulative, interpretive, and predictive possibilities that arguably transcend (at least in terms of efficiency and scale) those of any pre-digital technologies. The ways in which information is interpreted and manipulated through the digital medium can certainly be interpreted as forms of recording. In this chapter, however, I will argue that our uses of the computer to filter, analyze, and make sense of recorded information are better understood as specific forms of (technically mediated) "doing."

The growing diffusion of digital mediation in social practices in the past four decades brought reflections, discussions, and hazards concerning the possibilities and effects of such mediation to the fore of a vast and interconnected number of cultural contexts, from theoretical philosophy, to lawmaking, to the very activities involved in developing digital technologies. In that context, I will propose an understanding of "doing in the digital" that focuses on the roles and activities that the digital medium frames for its users. In particular, I will discuss digital environments (accessed through video games and digital simulations) as technical mediators that disclose new possibilities and new "grounds" for the discipline of philosophy. In other words, I will focus on the computer in its capability for shaping our thoughts and our behaviors in contexts that are conceptually and experientially more flexible than our everyday relationship with the world that we share as biological organisms (Gualeni 2015, xv).

I will commence this pursuit by clarifying that with "digital environment" I mean an artificial context upheld by digital technologies that permit specific kinds of experiences. These experiences are virtual,[2] and are characterized by varying degrees of perceptual permanence, mechanical consistency, and intelligibility (Gualeni 2015, 6–7; Chalmers 2016). Adopting this stance, artifacts such as an interactive database, a computer platform for the simulation of surgical procedures, a dungeon in an action-adventure video game, or the text editor I am presently working with can all be categorized as "digital environments" that are populated by "digital objects." (ibid.)

2   "Doing" in the Digital

Until this point, this chapter has outlined an understanding of the computer as a multimodal and multistable (see Note 1) technological medium that discloses various possibilities for "doing." As already outlined, some of those possibilities "reproduce and replace" those of previous media forms, while others can be recognized as specific ways in which the digital medium records, organizes, and reveals information and experiences. Different digital environments in different socio-cultural contexts can reveal activities through which we can interactively express ourselves, and new environments where we can relate to one another, pursue self-reflection, entertain ourselves, and so on.

If we are ready to accept these theoretical premises, then we also need to be open to embracing the idea that no single definition of "doing" can reasonably be expected to exhaustively capture the entire horizon of configurative and prefigurative practices that can be pursued and upheld in digital environments. In light of these observations, I cannot aspire to offer a complete analytical definition of "doing in the digital." Instead, what I propose is a working and inevitably incomplete taxonomy of future-oriented activities that the computer enables, frames, and translates.

The perspective from which I propose to understand "doing in the digital" relies on identifying two fundamental meanings that can be ascribed to the verb "to do." In English dictionaries, the definitions corresponding to "to do" refer to a number of practical activities such as preparing for use, traversing, arranging, mimicking, producing, killing, completing, sufficing, behaving, managing, and many more. With the objective of using the verb "to do" in a way that is helpful and revealing in relation to the possibilities and effects of virtual experiences in digital environments, I propose to use the following, dual, categorization of "doing":

1. "Doing" understood as performing intentional actions (or "doing as acting").
2. "Doing" in the sense of designing or constructing something (or "doing as making").

This distinction is obviously hard to defend in terms of sharpness and rigour, as the two categories have areas of conceptual overlap. In that respect, my objectives for the upcoming sections are to clarify the use of this dual understanding (Sects. 2.1 and 2.2) and to address the ambiguities and limitations inherent in taking this foundational methodological decision (Sect. 2.3). In simpler words, I will offer a detailed and practice-oriented account for each of the two proposed understanding of "doing," and will critically reflect on some of the ambiguities inherent in using this dual framework.

2.1   On "Doing as Acting"

Unlike the relationships that can be established with pre-digital technologies of communication and representation, I argued elsewhere that digital environments can disclose experiences, and not mere representations (see Gualeni 2015). These experiences, as already specified, are characterized by varying degrees of perceptual permanence, mechanical consistency, and intelligibility. Additionally, they often require inputs from their users-players. The latter are, thus, commonly understood not only as active participants, but also as active producers of the contents of the experience (Aarseth 1997; Calleja 2011, 55).

With "doing as acting" I indicate various types of experiences within a digital environment. These experiences are recognized as emerging from the users taking intentional actions within the boundaries set by the constitutive rules and the mechanical limitations of a particular environment. The "activity horizon" disclosed by a digital environment thus constitutes the material and conceptual context for "doing as acting" within it.

In line with the work of psychologist James J. Gibson, in this chapter I use the term "affordances" to indicate the possibilities of action offered by an object or an environment (1979). Understood in this way, affordances invite or even frame ways of understanding environments and situations and possible behaviors within them. As for any other forms of technological mediation, digital environments present experiential and conceptual affordances in a number of ways: from materially disclosing certain possibilities of use, to inviting or rewarding specific interactions by means of aesthetic stimuli. Adopting this perspective, we can say that "doing as acting" in a digital environment refers to the activity of operating with, within, and even against the possibilities of action it offers (its affordances). For the implied readers of this text, I imagine that activities such as playing a video game or sending messages through an e-mail client are relatable examples of "doing as acting."

"Doing as acting" within digital environments can function as factors of socio-cultural change in many, and often concurrent, ways. In relation to the objectives of this chapter, I consider the following to be among the most significant ways in which the digital medium invites and reveals ways of "doing as acting":

- in its offering the possibility to pursue professional training, for example in the aviation or medical fields, allowing people to attain a degree of practical expertise without the risk of causing actual damage to people or equipment;
- by exposing media users to a "procedural" form of rhetoric; that is to say, a form of persuasion that takes place through rule-based representations and interactions (Bogost 2007, ix);
- in its disclosing artificial experiential settings where various and innovative forms of psychotherapeutic activities, from behavioral conditioning to the treatment of anxiety disorders, can be pursued (Riva et al. 2016);
- by inviting actions in safe digital environments that elicit feelings of ownership, agency, and personal investment in relation to learning scenarios, making computers efficient and engaging educational technologies to be used in and out of schools (Gee 2007);

- in its granting interactive access to virtual contexts where arguments, ideas, world-views, and thought-experiments can be shaped, explored, manipulated, and communicated (Gualeni 2015, xvi).

2.2   On "Doing as Making"

Not all forms of "doing in the digital" consist in operating within the existing affordances of a digital environment. There are practices that rely, instead, on adding, removing, shaping, and re-shaping those very affordances. In other words, with "doing as making" I indicate activities that are involved in framing "possibilities of action" within digital environments.

In Sect. 2.1, I briefly discussed the use of video games and e-mail clients as examples of "doing as acting." For the sake of consistency and style, I am going to refer to the same two groups of applications to illustrate the second understanding of "doing" in digital environments.

Video games and e-mail clients can be relatively intricate digital artifacts. Their technical development typically involves several interconnected activities that frame the users' "possibilities of action" within their respective digital environments. Among those defining activities, we can include the programming of certain functional behaviors and the design of methods for action and feedback (for example, the decisions concerning user interfaces, the size of its elements, the visual design of its various configurations, the ways they respond and adapt to user interaction, the mapping of controls over peripheral devices, and so on). Those activities exemplify a form of "doing as making" in digital environments, one in which the "designers" (those who "do by making") set up digital interactive "stages" for those who "do by acting" within them. In more general terms, I propose to understand "doing as making" as the intentional manipulation of affordances.

With the objective of clarifying how this second understanding of "doing" resonates to the objective of contributing to our understanding of the socio-cultural meanings of our progressively more involved relationships with the digital medium, I believe it is helpful to introduce the concept of "building" as introduced by philosopher Davis Baird. In Baird's view, building (doing, constructing as a heuristic practice) offers an opportunity to correct the bias in the humanities which binds its methods and outputs exclusively to language (Baird 2004). According to this view, we should be open to pursuing and communicating scholarship through designed artifacts, whether digital or not. Baird's perspective participates in the idea that language is ill equipped to deal with entire classes of knowledge that participate in humanistic inquiry (Ramsay and Rockwell 2012, 78). Following Baird, Bogost similarly discusses the activity of constructing artifacts as a viable and much neglected philosophical practice that "entails making things that explain how things make their world" (Bogost 2012, 93). Neither Baird nor Bogost exclude digital artifacts from their approaches. On the contrary, in his work as a game designer, Bogost openly addresses the possibilities to use digital environments (or video-game worlds) for persuasion and cultural change.

In a way which is comparable with the two perspectives outlined in the previous paragraph, my wider approach to the digital medium (rooted in philosophy of technology and game studies) embraces digital environments as the contexts where new ways of pursuing the humanities have already begun to arise. More specifically, my work focuses on the experiential effects of digital environments (disclosed by "doing as acting" within them) and on the new possibilities for self-transformation as well as philosophical inquiry that become available while designing digital environments (emerging from "doing as making") (Gualeni 2014b, 2016).

Playing in counterpoint with the theoretical and interpretative approaches adopted by other chapters in this book, this chapter not only proposes an understanding of the digital medium that focuses on its disclosing various forms of "doing," but also shows how this *praxis*-oriented approach to digital mediation can be put into practice. In the second half of this chapter, the proposed dual perspective on "doing in the digital" will be expanded upon and concretely exemplified by my hands-on work as a game designer. More specifically, in Section 3, I will analyze the design of *Something Something Soup Something*,[3] an "interactive thought experiment" that I developed in 2017 with the support of the Institute of Digital Games (University of Malta) and *Maltco Lotteries*.

The reason why it is relevant, if not crucial, that my arguments include and discuss practical applications should be evident to the reader. The choice of analyzing and discussing one recent work of mine in particular might be, instead, less clear-cut. I will focus on *Something Something Soup Something* as having worked on the design and the development of that video game (if we agree to call it such) offers what I consider to be definite advantages. My hands-on involvement with the project granted first-hand insights concerning how the "players" responded to the experience of "doing as acting" within its digital environment and what they took away from the experience. It also allows me to present and discuss the constraints, the contextual and technical factors that framed that specific instance of "doing by making." The design process of *Something Something Soup Something* as well as its cultural strategies and aspirations will be presented as deliberately leveraging both dimensions of "doing in the digital" for philosophical purposes.

Before that, as anticipated, I deem it necessary to propose some critical methodological reflections on the ambiguity and the shortcomings of having embraced this dual understanding of "doing" as the theoretical framework for this chapter. By offering those additional considerations, I am not merely pursuing the objective of academic rigor, but also that of clarifying the range of applicability of my argument.

2.3   A Tale of Two Doings

This chapter proposes an understanding of the digital medium that focuses on the computer's capability of disclosing experiential ways of comprehending and behaving. In this context, the verb "doing" was understood as having a dual meaning: "doing as acting" ("playing" with and within the affordances of a digital environment) and "doing as making" (designing and manipulating digital affordances, setting up virtual, experiential environments).

As noted before, the various analogies between the two categories often make it impossible to neatly separate the two categories. To further elaborate on this point, I will yet again resort to examples from a specific, and hopefully specifically familiar, declination of digital environments: that of video games. Several video games and video game genres offer their players the possibility to modify some of the video game's parameters, to modify their in-game avatar, to set up new environments for play, and more in general to produce interactive content. Given a playful digital environment that offers possibilities for the manipulation and creation of new content, are not its players engaging that specific video game also as "makers"? Or does the fact that those design activities are disclosed to the players as explicit affordances of the video game make them examples of "doing as acting" rather than "doing by making"?

Sharply separating the two categories of "doing" might be an even more daunting task if we consider "folk practices" related to video games, such as *modding*, where a community of players uses (or even develops) digital tools to modify the contents of an existing video game. Another practice that I find particularly relevant to examine in this line of inquiry is that of "playing with style" (Parker 2011). Players who "play with style" fashion their digital experiences according to specific, restrictive principles, and approach their inhabiting digital environments as occasions to "shape themselves" into specific ethical and/or aesthetic subjects. Activities identifiable as "playing with style" can be recognized at many overlapping levels of the experience of "gameplay," where by "gameplay" I mean "doing as acting" within the functional and conceptual affordances of a playful (digital) environment. The stylization of players' actions can be exemplified in acts of videoludic sportsmanship (as forms of ethical and aesthetic self-fashioning in competitive multiplayer games), or in the time and attention dedicated to the creation and the aesthetic refinement of a player's in-game avatars (instances of aesthetic self-fashioning in both single-player and multiplayer games). They can also be recognized in acts of resistance and rebellion to various ideologies when those ideologies are framed in game systems in the form of functional and aesthetic affordances. Often cited instances of such "critical" and "subversive" ways of playing are pacifist runs (consisting in playing survival, war, or adventure video games resorting to violence as little as possible or not at all) or vegan runs (playing video games without pursuing in-game actions that kill or hurt in-game animals, exploit in-game animals, or make use of in-game animal products) (Westerlaken 2017).

As just observed, while practicing in-game self-fashioning, players often restrict their gameplay possibilities to uphold certain aesthetic values or some aspects of narrative consistency that they consider important. They may refuse taking unrealistic or unethical routes in the pursuit of their in-game objectives. In general terms, "playing with style" consists in the players' voluntary addition of personally meaningful restrictions to their always-already constrained possibilities of action within a certain digital environment (Gualeni 2014b). Should we consider these approaches to "gameplay" as forms of "doing as acting"? Or should we interpret "playing with style" as a way to critically engage with the affordances of a certain digital environment, instead of adapting to them? And if so, would not "playing with style" appear as a way to "make oneself" in a context that is less inexhorable and serious than the world we inhabit and share as biological creatures?

With the objective of furthering and completing the exploration of these methodological ambiguities and limitations, it may also be useful to consider the perspectives of those who design digital affordances, that is to say those of the programmers and designers of digital environments. As should be obvious to anyone with a basic familiarity with computers, the design of applications and activities that takes place within the digital medium does not take place in a socio-technical *vacuum*. The time and money available to create a new piece of digital design are almost ubiquitous constraints, and so is the fact that cables, whirring fans, input devices, and electrical power are all still necessary, material components of any activity taking place within digital environments.

Evidently, a digital environment is always defined by affordances that are in place before any kinds of "doing in the digital" can occur. In relation to this point, it may be important to remind the reader that, by definition, all digital environments are upheld by digital technology, which imposes a very specific understanding of rationality, of time, space, and causation as a prerequisite for most forms of "doing" with and within it.[4] Willard McCarty similarly emphasized the inherent dependence of any computing system on an explicit delimited conception (or model) of the world (McCarty 2005, 21).

Additional affordances that are inherent in the digital medium can be identified in how programming languages and middleware invite or downright impose specific ways of understanding, storing, representing, and manipulating information. Digital actions that were earlier identified as instances of "doing as making" can, from this perspective, always be understood as forms of "acting" in the digital medium. Any uses of computers can, in fact, be recognized as being always already constrained by the very "digitalness" of computers: by the structure of the languages devised for programming computers, by conventions and ways of operating that are required by specific software applications, and so on.

At the beginning of this critical sub-section, I clarified that "doing as acting" in a digital environment does not necessarily exclude a number of activities that can be ascribed to the category of "doing as making." After that, I outlined the perspective of a designer of digital environments, for whom all forms of "doing in the digital" can be also understood as forms of "doing as acting." In light of these ambiguities, skepticism towards the viability and the usefulness of the proposed distinction between "doing as acting" and "doing as making" seems to be well motivated. Would it not be more fruitful not to attempt a formalization of "doing in the digital," and instead to embrace a less precise, less disjointed, and less problematic understanding of "doing"? In other words, what are the theoretical or practical advantages in setting up that distinction? Is it not merely leading to exceptions and complications?

Faced with these questions and difficulties, I present two sub-arguments that pursue the following objectives:

- explaining why I consider my theoretical arguments fruitfully contribute to our knowledge of the transformative effects (both at the and the cultural scale) of digital mediation in social practices;
- paving the way for the second, *praxis*-oriented sections of this text (Sects. 2 and 4).

My first sub-argument follows from what was observed in Sect. 2.2, where notions such as "building" (embraced as a heuristic practice) were presented as a group of interconnected activities that allow us to pursue and communicate scholarship through artifacts. The growing interest in pursuing cultural objectives through activities and artifacts is also particularly visible in scholarly fields that embrace practice-based research methodologies such as research through design or artistic research. However, those activities as well as the various possibilities offered by practices and artifacts in terms of discovery, experimentation, and communication, remain largely under-theorized. A practical approach to themes and questions that involve the development of artifacts, the prefiguration and the configuration of situations and alternative possibilities of action, is already a relatively common approach in the context of philosophical inquiry. This is particularly evident in branches of philosophy that are characterized by obvious fields of application such as philosophy of science, philosophy of design, and philosophy of technology. The notion that one can "do philosophy" in the sense of a practical experiential engagement with something is an explicit theme of my own work as well as in that of Dutch philosopher Clemens Driessen (Driessen 2014; Gualeni 2014a; Gualeni 2015; Westerlaken and Gualeni 2017).

Regardless of the progressive diffusion of scholarly approaches that involve various degrees of "doing," it is still remarkably rare to encounter a description of what "doing" means with regard to cultural or academic pursuits. Despite several points of discomfort in the theoretical stance that I propose in this chapter, and despite its limited applicability, I find the distinction between "doing as acting" and "doing as making" to be useful in facilitating the discussion concerning the design decisions and the experiential effects concerning certain artifacts, regardless of their actual, digital, or hybrid constitution. I will demonstrate this idea at work in the next section of my text through an analysis of my work as a "digital doer." Admittedly, the "facilitation" discussed above is only such in a very restricted portion of what is, instead, a much vaster horizon of "doing." On the upside, the distinction between the two activities allows us to discuss artifacts and experiences in relation to those artifacts in ways that are not obfuscated by all-encompassing approaches to what "design" is (or could be), or by less-examined approaches to "doing."

My second sub-argument as to why my dual approach to understanding "doing in the digital" focuses on the limited applicability of the proposed approach, a topic that was touched upon in the previous paragraph. Despite the ambiguities and difficulties with the offered distinction between "doing as acting" and "doing as making," I believe the approach can be viable and useful if we agree to limit its application to observations, experiences, and interactions with a certain artifact. What I mean to say here is that this approach to "doing" becomes viable only when disregarding the socio-cultural and technical factors that influenced the design and the development of the artifact in question. In the case of the set-up of a digital environment, this limitation is tantamount to focusing on the interactions, the experiences, and the operations that take place *within* that environment while ignoring the social norms, the economic conditions, the hardware limitations, and the software affordances that contributed to its design. In addition to that, and still as part of this second sub-argument, I should like to explain that I do not propose to understand "doing as acting" and "doing as making" as two separate and mutually exclusive practices, but rather as two active "roles" that an artifact can disclose for agents who are "doing" with and within it. Depending on the affordances offered by an artifact, these roles can be completely divorced, can co-exist, can overlap, and can even dynamically shift among all those configurations.

To summarize the two sub-arguments that were just presented, I first identified the main theoretical stance of this chapter (presenting a dual understanding of "doing") as a problematic one, but one that can nevertheless be considered valuable. Its value was recognized in the possibilities it offers for further developing our understanding of "doing" as a cultural practice. Following this initial part, I delineated the field of validity of the proposed theoretical stance, restricting its use to observations, experiences, and interactions with the artifact in question, or (in our case) within a certain digital environment. Finally, I clarified that I do not consider the two understandings of "doing" as two separated practices, but as dynamic active roles framed and facilitated by the functional and conceptual affordances offered by a certain artifact. On the basis of these premises and critical reflections, in the next section I will articulate a practical understanding of "doing in the digital" in ways that are exemplified by the design decisions and the gameplay affordances of *Something Something Soup Something*.

3   Philosophical Soup

The official website for *Something Something Soup Something* (http://soup.gua-le-ni.com) qualifies our work as an "interactive thought experiment"; that is, as a mental exercise designed with a philosophical scope. Mostly presented through the mediation of language, thought experiments traditionally consist in imagining hypothetical situations and thinking through their possibilities and consequences. The interactive experience of thought experiments with and within digital worlds allows us to overcome, I shall argue, some of the limitations and biases that are inherent to the representational and hermeneutic possibilities of language.

My claim, as will be specifically explained in the upcoming sections of this chapter, is valid for "doing as making" within digital worlds as well for "doing as acting"; that is to say, both as people whose thought processes are activated by mental exercises and as people who design thought experiments.

The website for *Something Something Soup Something* also summarizes the narrative context for our video game as follows:

> The year is 2078. It has been a decade since *Homo Sapiens-Sapiens* perfected teleportation technology. Rather than using teleporters to overcome scarcity or to oppose various forms of oppression, this new technology is used to further promote economic exploitation. We hire aliens to manufacture goods on their planets, and then beam the products of their cheap labour back to Terra. Due to categorical differences in language and cognition, however, the aliens often misunderstand what humans ask them to produce [. . .].

In *Something Something Soup Something*, the "player" experiences the game-world from a first person perspective: that of a kitchen worker in a restaurant on Terra. The interactions with the digital environment of *Something Something Soup Something* begin in a dimly lit, dirty stairwell leading downwards. The disembodied voice of the narrator informs the player that she is supposed to head to the kitchen and activate the teleporter machine. Once the teleporter is activated, the narrator continues, it will materialize the dishes ordered by the human customers of the restaurant. As mentioned earlier, however, the aliens who remotely produce those food items often do not have a clear idea of what we want and what we mean by words such as "soup." This fundamental equivocality will put the player in a puzzling situation.

Something Something Soup Something is a short, free video game that takes place in a game world composed of a stairwell leading to an underground kitchen and the kitchen itself. The player needs a mouse and a keyboard to interact with the digital environment and comply with his or her in-game duties as a kitchen worker.

The following two sub-sections of this chapter will discuss *Something Something Soup Something* as a digital artifact that leverages both dimensions of "doing in the digital" for philosophical purposes. This analysis will focus on the gameplay affordances that the video game offers and the game-design decisions that contributed to shaping its interactive argument. More specifically:

- Section 3.1 will concentrate on the digital environment of *Something Something Soup Something* as it can be experienced by a player; that is to say, on how the game interactively discloses meaning and information to someone "doing as acting" within it.
- Section 3.2 will instead focus on the video game from the point of view of its "makers," examining the design decisions taken during the conceptualization and the development of the game in relation to its scholarly objectives.

3.1   Playing with Digital Soups ("Doing as Acting")

With the objective of obtaining cheap soups from outer space, the player will enter the kitchen at the bottom of the stairwell. Looking for the teleporter machine to activate, the player has a chance to pick up additional details and information about the fictional world of *Something Something Soup Something*. The kitchen seems to be neglected and dirty. From the looks of it, it has been quite a long time since anybody cooked in there, let alone cleaned it. A bulky piece of machinery occupies the back end of the room; in all likelihood, the conspicuous greenish apparatus is the very teleporter that the player is looking for (see Fig. 1).

Fig. 1 An overview of the dystopian kitchen of *Something Something Soup Something*

Over a rusty sink, a flickering screen reports that there is an outstanding order for the kitchen consisting of 20 soups.

Initially, trying to activate the large greenish machine does not produce any visible results: if that weird contraption is indeed the teleporter machine, it seems to be inactive, or possibly broken. While most of the elements of the kitchen seem either unresponsive or out of service, one small and quirky device at the center of the room captures the player's attention. That strange piece of technology seems to be functioning and ready to work. A couple of cables connect it to the (presumed) teleporter. Its instruction leaflet qualifies the device as an *STJ 1062 M* signal decoder, where the "M" indicates that the cosmic noise cancellation for this device is to be performed manually.

Once turned on, the decoder automatically starts to receive a signal. In a way that is reminiscent of video game genres such as rhythm games or infinite runners, the player is prompted to interact with the decoder and to try and reduce the interference of the cosmic noise while the signal is being received (see Fig. 2).

Failing to keep the level of corruption of the signal under control triggers a reset mechanism that automatically turns the decoder device off. Managing to keep the cosmic noise to a minimum instead allows the players to successfully complete the transmission. Once the signal has been properly processed, the decoder automatically powers up the teleporter machine, and the narrator grimly comments on the successful operation. "The soups have been beamed in," he says, "but aliens often misunderstand what we ask them to produce. Which of the dishes in the teleporter are effectively soups? Which of them can be reasonably served as soups to the human customers?"

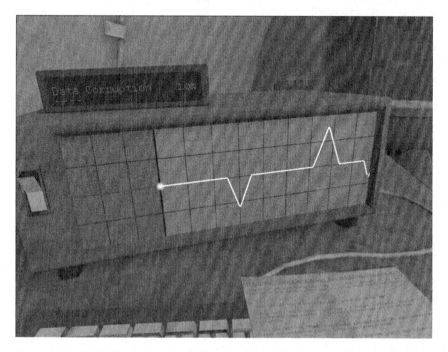

Fig. 2 The *STJ 1062 M* signal decoder receives a noisy transmission from outer space

Interacting with the teleporter machine, the players can now choose which dishes among the ones presented to them are viable soups for humans and which ones are not. In this phase of the game, *Something Something Soup Something* challenges the players to assess 20 different dishes defined by a number of different characteristics (constitutive ingredients, edibility, temperature, accompanying serving and eating tools, and so on).

The choice of whether each of the received items is a soup or not is performed for each individual dish by clicking one of two buttons on the teleporter itself. One button has the word "soup" printed on it, and its design features an upward-pointing symbol (hinting at the restaurant upstairs). The other button is characterized by the words "not soup" and a downward-pointing symbol (indicating the recycling chamber in the bottom part of the teleporter machine, see Figs. 3 and 4).

Among the players we observed and interviewed during the development of our interactive thought experiment, several asked us whether the bizarre and uncanny qualities of the supposed soups that appeared in

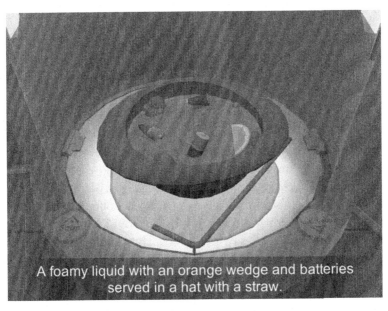

A foamy liquid with an orange wedge and batteries served in a hat with a straw.

Fig. 3 An example of a bizarre dish (possibly a soup) that was beamed in by the teleporter

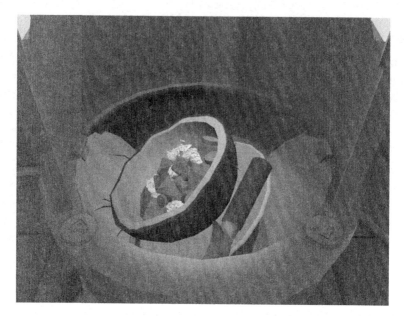

Fig. 4 This particular dish was discarded by the player. Too cold for a soup? Too solid, perhaps?

front of them were the product of their poor performance with the cosmic noise cancellation task. Had they performed better at reducing the cosmic noise, the players wondered, would they have received more acceptable, more conventional soups?

This is the first philosophical question that we hoped to trigger in the players in relation to their practical interactions with *Something Something Soup Something*: would a more precise communication bypass problems of conceptual ambiguity? Could other philosophical problems be solved through a rigorous reformation of our language? This point and its design implications will be discussed in finer detail in the following sub-section.

The second, and perhaps more central, philosophical question that we hoped to instill in the players by means of (playful?) digital activities and interactions was, clearly, whether we can exhaustively and clearly define what soup is. A minority of the observed players took the soup-selection task facetiously, and reasonably so: the video game takes place in what is obviously a fictional world, and the task itself has outlandish premises.

Most of the other people we observed and interviewed concerning their interactions with *Something Something Soup Something* reported having taken the selection procedure seriously. While choosing which dishes could be reasonably served to humans as soups, this second group of players quickly came to the realization that making definitive decisions about what soup is (or what soup is for them) is neither banal nor unproblematic. Some of these players reported having reworked their conceptualization of soup in the process of selecting them, as the activity made them progressively more sensitive to possible discriminating factors as well as more aware of the very choices that they were making.

After having assessed all the received dishes from outer space, that is a batch consisting of 20 possible soups, the player is presented with a synthetic summary of their soup-related decisions on the kitchen display (see Fig. 5). This conclusive part of the experience of *Something Something Soup Something* serves the obvious purpose of making the players' own understanding of soup into an object for their own, as well as other people's, critical evaluation.

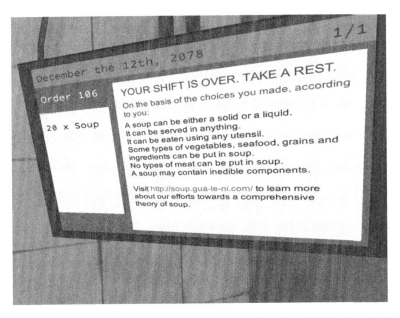

Fig. 5 The players' soup-related choices are summarized on the kitchen display

As indicated in the concluding summary on the kitchen display (see Fig. 5), the official website for *Something Something Soup Something* (http://soup.gua-le-ni.com) offers additional information concerning the video game, its design, and its philosophical aspirations. In this sense, the website functions as a complement to gameplay, disclosing and discussing aspects pertaining to the philosophical arguments of the latter in a more traditional (that is, more textual and linear) and explicit fashion.

3.2   Designing Digital Soups ("Doing as Making")

Section 2.3 (*A Tale of Two Doings*) briefly discussed the interactive experience of *Something Something Soup Something* as having the potential for disclosing a philosophical argument through gameplay. The present one will try to articulate instead the most salient design decisions for *Something Something Soup Something* in relation to the video game's intended meaning.

In the official webpage for the video game, we explain that *Something Something Soup Something* is designed to experientially and interactively disclose certain philosophical notions to the players. Its main philosophical point concerns the idea that even an ordinary, familiar concept like "soup" is vague, shifting, and impossible to define exhaustively. As already explained, the players explore this concept as they struggle to identify what a soup is through a series of binary choices. The synthetic summary of their soup-related choices that is presented on the kitchen display at the end of each game session further emphasizes the intended message of the game, explicitly confronting the players with their inevitably incomplete understanding of what soup is (see Fig. 5).

Section 3.1 also explained that the soup selection part of *Something Something Soup Something* follows a real-time mini-game. In this mini-game, the player is asked to manually reduce the cosmic noise interference in order to correctly receive a signal from outer space (that is, to download a batch of soups into the teleporter machine). This playful real-time activity was designed to complement the main philosophical point of the soup-selection experience. In allowing the player to refine and perfect the transmission of data, the mini-game seems to suggest that there could be a way of receiving less problematic, less ambiguous soups . . . And maybe even of accomplishing a clear and complete analytical definition for what soup is.

Does the game offer the possibility for perfectly sharpening communication? Did the designers take a position compatible with Austrian philosopher Ludwig Wittgenstein's early work, according to which philosophical problems arise from our misunderstanding the logic of language (Wittgenstein 1961)? And if so, did they materially inscribe in the functional affordances and aesthetic aspects of *Something Something Soup Something* the perspective according to which philosophical problems and conceptual difficulties are ultimately caused by illogical and inaccurate ways of communicating? Or is the video game aligning instead with the later work of Wittgenstein and his belief that language is inherently messy, defective, and always determined by its contextual use?

The noise cancellation mini-game is designed to be impossible to execute flawlessly, meaning that despite the best intentions and the most attentive efforts on the player's part, data reception cannot ever be completely accomplished. In other words, our design aligned with the later Wittgenstein, that of the *Philosophical Investigations*, and was guided by the concepts of "language games" and "family resemblances" that the Austrian philosopher articulated in that text (Wittgenstein 1986). Provided that the noise cancellation is sufficient for the reception of the signal (that is, data corruption remains below 100%), then the players' efficiency in reducing the cosmic noise is irrelevant. The bizarre dishes presented in our "thought experiment in a digital world" do not depend either in their order or composition on how accurately cosmic noise was reduced. Soups are, instead, generated by a fairly simple algorithm designed to ensure that the players encounter every possible "soup feature"[5] at least once in each individual game session.

The conceptual disconnection between the noise cancellation task and the soup selection one is not obvious, and is never explicitly revealed to the player. One might only start to realize how those two aspects of the games work in relation to one another only after several play sessions. Given the relative obscurity of this "interactive corollary" to the main philosophical theme of *Something Something Soup Something*, one might raise the question of whether it was an efficient design decision to include the decoding mini-game to begin with. In response to this likely interrogative, I will introduce three arguments in support of why I consider the noise cancellation task to be a meaningful and desirable part of the gameplay of *Something Something Soup Something*:

1. As already discussed, the mini-game hopefully raises questions concerning the relationship between the signal from outer space and the qualities of the received soups. Through gameplay, players may discover that there is no obvious connection between the two phenomena beyond one being a simple on/off mechanism for the other. Regardless of how sharp and clear the players manage to make the signal, the soups' composition will remain odd and ambiguous. Metaphors aside, what the game is communicating here is that despite our attempts to reform language and sharpen communication processes, we still live and act in a world where the meanings we attribute to words are inherently indefinite and constantly shifting. We found this to be an important corollary to our main interactive philosophical argument, and we did not feel that the relative subtlety of this "interactive argument" was a valid enough reason to exclude it from the experience of gameplay.
2. The decoder mini-game contributes to making the embedded narrative to the game richer. Both its rickety design and the crude information contained in the instruction manual add further depth and thematic detail to the squalid game-world of *Something Something Soup Something*. Additional aesthetic and narrative detail in thought experiments (regardless of their literary, cinematic, or virtual mediation) has the potential to further stimulate the imagination of the recipient, which can also serve the philosophical purposes of the thought experiment itself (see Gendler 1998; Davies 2008, 10–11).
3. The traditionally ludological structure of the mini-game and its focus on the quantification of player performance was conducive to a third philosophical theme that we tried to weave into the video game and that was outlined on the official website of *Something Something Soup Something* as follows:

Something Something Soup Something is also designed to stimulate reflection on the possibility to analytically define what a game is. Our "interactive thought experiment" involves some narrative, it allows for a bit of exploration, it features a section that quantifies the performance of the "players" in relation to a pre-established goal [. . .]. But does the presence of those "ludological ingredients" warrant its definition as a video game? What if only a part of it could be formally recognized as a video game? Is it even wise or productive to strive for a complete theoretical understanding of concepts like "soup" or "game"?

I find this point to be especially beneficial in contexts like introductory courses in game design or game studies. Either presented in class or assigned as playable homework, the experience of our short, free video game can become the occasion of discussions concerning the limits of formalism in analyzing and designing video games and playful systems more in general.

Until now, the discussion and analysis of the philosophical themes of *Something Something Soup Something* kept the activities recognized as "doing as acting" (where experiences and arguments emerge from the experience of gameplay) tidily separated from those that, instead, could be identified as "doing as making" (activities concerned with building digital environments and taking design decisions). Having spent a considerable portion of this chapter to stress the difficulties in neatly splitting these "roles," it would perhaps be confusing (or at least methodologically incomplete) if I did not take some time also to discuss the "gray areas" of *Something Something Soup Something*. In other words, for the sake of clarity and completeness, I consider it important to consider those moments of gameplay in which the players are also "makers," and those aspects of game development in which (given the limitations posed by the digital technologies used) the designers can also be recognized as "acting" within the constraints of digital environments.

In our small, dystopian game-world, the players (those who, once again, "do by acting") have admittedly very little space to express themselves or to deviate from the pre-determined tracks of the game and its narrative. *Something Something Soup Something* is set in an inescapable kitchen in which the players are supposed to behave like employees, a loosely defined identity that they are not allowed to abandon as long as they act in that specific digital environment. Players can however "style" their in-game activities in a number of ways, most of which involve sabotaging their in-game identity as employees. In other words, the players are not given opportunities to take the role of "makers" apart from their possibility to "style themselves" rebelliously against the video game's implied ideology and expected gameplay. Such subversive acts could be exemplified as deciding to not comply with the fictional role of a kitchen worker or to make random or nonsensical choices in relation to the proposed noise-suppression and soup-categorization tasks.

Taking the perspective of the designers (those who did *Something Something Soup Something* as makers), we need to acknowledge that their role of "makers" cannot be unambiguously separated from their "acting" within one or more digital environments. In the specific case of *Something Something Soup Something*, the game engine we chose, the 3D modeling software we used, and the computer platforms that we expected the video game would be played on (together with their respective technical specifications) had a formative influence on how several technical and conceptual game design decisions were made. The time and money allotted for the project were also influential in the development of *Something Something Soup Something* as they contributed to determine the amount of game elements (such as soup features) and game functionalities that could be included in the video game.

As should now be obvious to the reader, the philosophical themes developed and materialized in *Something Something Soup Something* were not uniquely shaped by our understanding of the work of Ludwig Wittgenstein or by our sensitivities and design prowess. In a way that is not dissimilar to the expressive and conceptual restrictions that its traditional written form imposed on Western thought, "doing" philosophy with the digital medium is constrained and "shaped" by distinctive technical, contextual, and interactive limitations.

What I would like to clarify in these concluding paragraphs is that I do not consider "doing in the digital" as an exceptional (or even as a particularly desirable) way of pursuing humanistic inquiry, nor do I consider the computer to be the ultimate philosophical tool. The use of digital environments and other forms of practical involvement with research and education can overcome some of the inadequacies and limitations that are inherent to an exclusively linguistic (or more specifically textual) mediation of thought. It is, however, a form of "overcoming" that inevitably brings about new problems, limitations, and discontents. The embedding of video games and computer simulations in social and cultural practices (philosophy being one of them) might, thus, best be pursued on the basis of the understanding that, as with any other form of mediation, their virtual worlds disclose reality in specific ways, and that such ways are always inherently both revealing and concealing (Gualeni 2015, 94).

4   Conclusion

This chapter purposefully played in counterpoint with the general theoretical orientation of the book of which it is part. Instead of focusing its attention on the recording and archiving capabilities of the digital medium, it proposed an understanding of the digital medium that focused on its disclosing various forms of "doing."

On the basis of a theoretical framework rooted in media studies and game studies, this chapter focused on digital forms of "doing." The uses and advantages of the methodological separation between "doing as acting" and "doing as making" in virtual worlds were theoretically articulated in Sect. 2 and then discussed in terms of their practical results and effects in Sect. 3. In the third section, a playful digital world that I recently designed was specifically analyzed as leveraging both dimensions of "doing in the digital" for philosophical purposes.

Traditional scholarly approaches chiefly (if not exclusively) present insights, analyses, interpretations, as well as new perspective in textual form. In an attempt to overcome and complement the exclusively linguistic approach to the humanities (and philosophy in particular), I advocated for a compromissory and multi-media-driven approach to the development, negotiation, and dissemination of notions and arguments.

In concluding this essay, allow me to explicitly add that I am not proposing to understand the digital medium exclusively as a medium for "doing." I consider "doing in the digital" to be one of the current dominant uses of the computer. I want to clarify, however, that I am well aware that it would be limiting and perhaps counterproductive to embrace "doing" as the sole use of the digital medium. Other activities, such as recording, reading, discussing, and listening, can all coexist and collaborate with "doing" in developing content and shaping culture in an number of ways. In this sense, the "soup" analogy might also be fitting for our interpretation of the meanings and effects of digital mediation. Similar to the conceptual definition for what a soup is, developing a coherent and complete understanding of the possibilities and effects of our digital activities is a messy task, one that heavily depends on cultural and historical factors, and one that, notwithstanding our categorization efforts, may remain impossible to capture exhaustively.

Acknowledgments Several, recurring conversations with Michelle Westerlaken in the past months were both inspiring and shaping for the arguments offered in this chapter. If this text adds anything interesting or useful to the current discourse, she is probably the one to blame for that. In this acknowledgments section I would also like to thank Johnathan Harrington and Isabelle Kniestedt, who significantly contributed to the design and the development of *Something Something Soup Something*, to the preliminary research and technical work leading to it, and to the editing of this chapter.

Notes

1. The terms used to indicate this aspect of flexibility in our relationship with technologies are "multistability" (as introduced by Don Ihde) or "metastability" (coined by Bernard Stiegler, who adapted the concept from Gilbert Simondon). Both notions indicate, in different ways, that the relationships that people establish with technologies and the meanings that they attribute to such relationships are always influenced by a variety of socio-technical factors and situations.

2. The adjective "virtual" was originally used in modern Latin to encapsulate the idea of "potentiality." *Virtualis* is a late-medieval neologism, the existence of which became necessary when Aristotle's concept of δύναμις (*dynamis*: potentiality, power) had to be translated into Latin (Van

Binsbergen 1997, 9). The concept of 'potentiality' at the etymological foundation of the adjective "virtual" provides the background for understanding why, at least in one of its interpretations, it is used to indicate the latency of certain possibilities inherent in a specific artifact, combination of artifacts, or state of things (Gualeni 2015, 54–5). A more common connotation of the adjective "virtual" was presented by Pierre Lévy, a connotation that did not stand in opposition to "actual" in the sense discussed above, but to "actual" in the specific sense of something that is pertinent to the world humans are native to (Lévy 1998, 14). The latter understanding of "virtual" has definite affinities with a digitalist approach to virtual reality such as the one heralded by David Chalmers (Chalmers 2016, 2, 3)

3. *Something Something Soup Something* is available for free for a number of different computer platforms at http://soup.gua-le-ni.com. It was designed and written by Stefano Gualeni, in collaboration with Isabelle Kniestedt (art and programming), Johnathan Harrington (field research and additional design), Marcello Gómez Maureira (web-design and additional programming), Riccardo Fassone (music and sound effects), Jasper Schellekens (narrator and research support).

4. See Chapter "8.1.2 Humans who calculate" in *Virtual Worlds as Philosophical Tools* (Gualeni 2015, 156–7).

5. During the development of *Something Something Soup Something*, we had to identify the properties and features that different people in a variety of different cultures use to describe soup. By doing so, we tried to remove our personal biases about what soup is (or is not) from the conceptual design of the video game. Inspired by Eleanor Rosch and Carolyn B. Mervis's linguistic experiments, we organised focus groups in different countries (Rosch and Mervis 1975). The various activities involved in those focus groups ensured that our conceptions of soup, as designers, were as inclusive as possible (Harrington 2017).

References

Aarseth, Espen. 1997. *Cybertext. Perspectives on Ergodic Literature*. Baltimore: The John Hopkins University Press.

Baird, Davis. 2004. *Thing Knowledge. A Philosophy of Scientific Instruments*. Berkeley: University of California Press.

van Binsbergen, Wim. 1997. *Virtuality as a Key Concept in the Study of Globalisation*. Den Haag: WOTRO.

Bogost, Ian. 2007. *Persuasive Games: The Expressive Power of Videogames*. Cambridge: The MIT Press.

———. 2012. *Alien Phenomenology, or What It's Like to Be a Thing*. Minneapolis: University of Minnesota Press.

Bolter, Jay, and Richard Grusin. 2000. *Remediation. Understanding New Media*. Cambridge, MA: The MIT Press.

Calleja, Gordon. 2011. *In-Game. From Immersion to Incorporation*. Cambridge, MA: The MIT Press.

Chalmers, David. 2016. The Virtual and the Real. *Disputatio* (forthcoming). Accessed December 3, 2017. http://consc.net/papers/virtual.pdf

Davies, David. 2008. Can Film Be a Philosophical Medium? *Postgraduate Journal of Aesthetics* 5 (1): 1–20.

Driessen, Clements. 2014. *Animal Deliberation. The Co-Evolution of Technology and Ethics on the Farm*. PhD diss., Wageningen University.

Gee, James Paul. 2007. *Good Video Games and Good Learning. Collected Essays on Video Games, Learning and Literacy*. New York: Peter Lang Publishing.

Gendler, Tamar Szabó. 1998. Galileo and the Indispensability of Thought Experiments. *British Journal for the Philosophy of Science* 49: 397–424.

Gualeni, Stefano. 2014a. Augmented Ontologies; or, How to Philosophize with a Digital Hammer. *Philosophy & Technology* 27 (2): 177–199.

———. 2014b. Freer than We Think: Game Design as a Liberation Practice. In *Proceedings of the 2014 Philosophy of Computer Games Conference in Istanbul, Turkey*, November 12–15, 2014. http://gamephilosophy2014.org/wp-content/uploads/2014/11/Stefano-Gualeni-2014.-Freer-than-We-Think.-PCG2014.pdf

———. 2015. *Virtual Worlds as Philosophical Tools*. Basingstoke: Palgrave Macmillan.

———. 2016. Self-Reflexive Video Games: Observations and Corollaries on Virtual Worlds as Philosophical Artifacts. *G|A|M|E—The Italian Journal of Game Studies* 5 (1). https://www.gamejournal.it/gualeni-self-reflexive-video-games/.

Harrington, Johnathan. 2017. Something Something Game Something. In *Proceedings of The Philosophy of Computer Games Conference*, Krakow, Poland.

Lévy, Pierre. 1998. *Qu'est-ce que le virtuel?* Paris: La Découverte.

Manovich, Lev. 2001. *The Language of New Media*. Cambridge, MA: The MIT Press.

———. 2013. *Software Takes Command*. London: Bloomsbury.

McCarty, Willard. 2005. *Humanities Computing*. New York: Palgrave.

Murray, Janet. 1998. *Hamlet on the Holodeck: The Future of Narrative in Cyberspace*. Cambridge, MA: The MIT Press.

———. 2003. Inventing the Medium. In *The New Media Reader*, ed. Noah Wardrip-Fruin and Nick Montfort, 3–5. Cambridge, MA: The MIT Press.

Parker, Felan. 2011. In the Domain of Optional Rules: Foucault's Aesthetic Self-Fashioning and Expansive Gameplay. In *Proceedings of the 2011 Philosophy of Computer Games Conference, Panteion University of Athens, Greece*, April 6–9, 2011. https://gameconference2011.files.wordpress.com/2010/10/pocg_foucault_paper_draft3.pdf

Ramsay, Stephan, and Geoffrey Rockwell. 2012. Developing Things: Notes Towards an Epistemology of Building in the Digital Humanities. In *Debates in the Digital Humanities*, ed. Matthew K. Gold, 75–84. Minneapolis: The University of Minnesota Press.

Riva, Giuseppe, Brenda K. Wiederhold, and Andrea Gaggioli. 2016. Being Different: The Transformative Potential of Virtual Reality. *Annual Review of Cybertherapy and Telemedicine* 14: 3–6.

Rosch, Eleanor, and Carolyn B. Mervis. 1975. Family Resemblances: Studies in the Internal Structure of Categories. *Cognitive Psychology* 7 (4): 573–605.

Verbeek, Peter Paul. 2005. *What Things Do. Philosophical Reflections on Technology, Agency, and Design*. University Park, PA: Penn State University Press.

Westerlaken, Michelle. 2017. Aesthetic Self-Fashioning in Action: Zelda's Vegan Run. In *Proceedings of the 2017 Philosophy of Computer Games conference in Krakow*, Poland, 2017.

Westerlaken, Michelle, and Stefano Gualeni. 2017. A Dialogue Concerning 'Doing Philosophy with and within Computer Games'—or: Twenty Rainy Minutes in Krakow. In Proceedings of the *2017 Philosophy of Computer Games Conference in Krakow*, Poland, 2017.

Wittgenstein, Ludwig. 1961. *Tractatus Logico-Philosophicus*. New York: Humanities Press.

———. 1986. *Philosophical Investigations*. Oxford: Blackwell Publishing.

From Registration to *Emagination*

Alberto Romele

1 Three Paradigms Through a Bizarre Pair of Glasses

While this book is primarily devoted to the topic of registration, recording, and keeping track in the digital age, this chapter must be seen as a coda, in which I intend to lead the reader towards a series of reflections on imagination or, as I call it, *e*magination. For me, *e*magination does not contradict registration. It can rather be seen as an emerging property of the latter, just as registration can be seen as an emerging property of information and communication.

In order to introduce the topic, I will use a bizarre pair of glasses: the thought of a mostly unknown French author, Robert Estivals, who wrote a *General Theory of Schematization* (*Théorie générale de la schematisation*, GTS), published in three volumes between 2001 and 2003. This work

A. Romele (✉)
Université Catholique de Lille, Lille, France

Universidade do Porto, Porto, Portugal

© The Author(s) 2018
A. Romele, E. Terrone (eds.), *Towards a Philosophy of Digital Media*,
https://doi.org/10.1007/978-3-319-75759-9_13

went almost completely unnoticed. And even I may have overlooked it, had it not been for the intersection of two fortuities. The first one was being invited as a Visiting Professor at the MICA (Mediations, Information, Communication & Arts) Lab of the University of Bordeaux Montaigne. Estivals taught information science at that university between 1968 and 1993. Incidentally, he was not just an academic, but also an artist; he founded several avant-garde movements. In 1996 he created a museum in his hometown, Noyers-sur-Serein, called *La Maison du schématisme*. Today, some members of MICA Lab are trying to rehabilitate his work, along with that of his colleague Robert Escarpit. I happened then to attend a couple of sessions of the seminar "Archaeology of the Sciences of Writing" organized by Franck Cormerais in winter semester 2016/2017. The second fortuity was finding the three volumes of the GTS, used but in good condition, in a bookshop in the Latin Quarter in Paris, the same Saturday I came back from Bordeaux. And now that I have finished narrating the anecdote, I will tell you what I found interesting in this work.

First, Estivals distinguishes among three phases in the field of communication studies or, to be more precise, among three phases "in the evolution of the canonical schemas of communication." The first one, between Saussure and the 1960s, used to resort to an interpersonal schema, recognizing in this way the specific field of "communicology": the relation between entities. Let us think back, for instance, to the famous schema in Saussure's *Course in General Linguistic*, in which two persons discuss, and something seems to happen in their minds; and to its geometric version, in which it is clear that what is happening in these minds, in the coming and going of audition and phonation, is the relation between the acoustic image and the concept. According to Estivals (2003, 46), the archetypes of this schema are the line and the feedback. The canonical schema of communication is both linear and circular, as far as it privileges the enunciation, on the one hand, but on the other hand it also implies the possibility for the receiver to retake the initiative. Another schema of this kind is Shannon and Weaver's model of communication in which new concepts are introduced: information source, message, transmitter, signal, and so on.

In the second phase, between the 1960s and the 1990s, the debate shifted from interpersonal to social communication. The archetypes of the line and the feedback gave way to that of the network. Among others, Estivals proposes a schema by the electrical engineer Abraham Moles, who was also Professor of sociology and social psychology at the University of Strasbourg, and is today considered one of the pioneers of communication studies in France. In the third and last phase, that began in the 1990s, the concern of communication studies for linguistic, information engineering and sociology declined, and cognitive sciences took the upper hand: neurology, psychology, logic, and especially artificial intelligence (AI). The archetype, in this case, is Franck Rosenblatt's perceptron, created in 1958, that Estivals (2003, 64) also calls "networked machines," whose history of decay and revenge has been recently resumed by Domingos (2015).

This periodization has strong analogies with the one that characterizes the more recent history of digital media, and that has been presented in the introduction to this book. Moreover, the shift from social networks to neural networks Estivals speaks about is very close to the transition from recording and registration to *emagination* I am going to debate. The classic information theory cannot account for the digital and its current impact on society. What is at stake today is not so much information circulation, but the fact that all flow of information, and more generally all digitally mediated or accompanied action, leaves a voluntary, involuntary or induced digital trace. The notion of digital trace has been introduced in communication studies in France via phenomenology and semiotics, and has been used in English mostly by French authors, for instance Latour (2007). Such a concept has at least three sorts of advantages. First, it allows understanding digital registration and recording in the light of the history of "statistical reasoning" (Desrosières 2002) and "documentality" (Ferraris 2012). At the beginning of the twentieth century, to leave traces like documents or monuments was a privilege of the cultural and urban elites, or a condemnation for the excluded, imprisoned or stigmatized people. Still at the end of the last century, this phenomenon concerned only a few social acts such as christenings, marriages, divorces, and fines. Today, thanks to the digital, especially mobile and wearable, technologies, and through the development of what Jacek Smolicki calls in this book the "capture culture," traceability has become a "total social fact." Second, the concept of digital trace sends us back to the issue of the materiality of the support on which and through which the trace has been recorded. Third, the same notion reminds us of the interpretative rather than apodictic character of all disciplines related to it. The Italian historian Carlo Ginzburg (1989) famously theorized the existence of an "evidential paradigm" for human and social sciences, based on the interpretation and understanding of "clues": *spie* in Italian, *traces* in the French version of the same article.

Now, my hypothesis is that we are entering a third phase, in which not just information and registration are at stake, but also *e*magination. And this is the second reason that has made me appreciate the works of Robert Estivals. In the GTS, he argues that communication has to do with schematization, and this is the specific cognitive task Kant attributed to imagination in the first *Critique*. Estivals (2003, 80) defines schematization as the "whole of the processes of cognition and symbolization based on two complementary principles: the reduction and the organization of information, the arborescent and the reticular schematization." He also affirms that "everything happens as if the first act of human understanding is aimed at the reduction of information … Understanding is first of all reducing and isolating. … However, once the unity is achieved, the thought is led to involve concepts in relation" (Estivals 2003, 81). The example he proposes is that of working with a computer: looking for and selecting information, on the one hand, and constituting a corpus, a hypertext or a thesaurus, on the other hand. He also distinguishes between scheme, the operation of schematization in the human minds, and schema, which is its external manifestation, and can be linguistic, phrasal, textual, narrative or meta-textual. For him, communication is a series of relations of schematization, the scheme in the sender, the schema, which is its externalization, and the schema in the receiver. To tell you the truth, I am oversimplifying Estivals's theory slightly: on the basis of Piaget's work, he distinguishes, indeed, in every individual, between an intuition scheme and a scheme of the cognitive structure. Moreover, at the level of the schema, he introduces a series of passages and distinctions between linguistic schema, transcoding, graphic schema, gestural schema, and so on. In any case, I consider his approach interesting, because I also believe that communication and interpretation are forms of schematization, that is of simplification and articulation of recorded information about reality.

For me, hermeneutics, the art of interpretation, is not an exception, namely giving up habits when a lack of understanding occurs. It is rather always-already at work in my being-in-the-world. I cannot but recommend the reader to refer back to the works of authors such as Makkreel (1990) and Lenk (1995) on this point.

The limits of the GTS are, however, clear. First, there is an unjustified priority that is attributed to the linguistic schemas, while it has been noticed, for instance, that also gestures can be considered as tools for the synthesis that characterizes schematism and hence imagination (Maddalena 2015). Secondly, and more importantly, can we really say that schemas are mere externalizations of internal schemes? In my opinion, it is rather the opposite, in the sense that schemes can be seen as internalizations of the external schemas; or, to be less extreme, our schematizations often depend, with different possible degrees, on those syntheses that happen "out of our heads." The literature that helps me in supporting my thesis is vast, and comes from different schools of thought and approaches such as those of Clark and Chalmers's "extended mind," Stiegler's "tertiary retention," and Jack Goody's "graphic reason," recently adapted to computer sciences by Bruno Bachimont (2010), who introduced the notion of "computational reason." But I would like to focus, for a moment, on the perspective of a philosopher who sank into oblivion for a long time, and is now enjoying a great comeback: Gilbert Simondon, who specifically devoted a seminar to imagination and invention in 1965–1966 (Simondon 2014).

Simondon is one of the rare French thinkers who did not follow the linguistic turn in those years. He understands imagination as both externalized and technicized. According to Vincent Beaubois (2015, np), Simondon's imagination can be defined as practical and technical schematism. Indeed, the scheme is not a mental entity for him. It is rather an operation that is made in and by the things themselves. Our imagination consists in the capacity of perceiving some qualities in the things that are neither directly sensory nor are entirely geometrical. This is precisely the intermediary level of the schemes (here, of course, there is no distinction anymore between scheme and schema), which are in-between matter and forms. In other words, human imagination is the intuition of the schematization that happens in the things themselves. For human beings, invention is the possibility of establishing new analogies among different technological lineages. There is, then, not just passive intuition but also a procedural activity in human imagination. Such an activity, Beaubois points out, is not merely conceptual, but directly depends on the practical and even emotional frequentation of the technical objects. What is interesting for the rest of this chapter is also the fact that if invention is the capacity of finding out formerly unseen analogies in the technical world already at our disposal, then imagination has less to do with pure invention than with recombination. According to Simondon (1989, 74), "the inventor does not proceed *ex nihilo* … but from already technical elements." Similarly, in the "circle of the images" he proposes at the beginning of the 1965–1966 seminar, invention itself is presented not as pure creation, but as reorganization of the system of images (the term "image" stands often in this seminar for "scheme") which is already in our possession.

2   We Have Never Been Engineers

The case of digital technologies and media, or at least an emergent part of them, is particular. They do not just support or mediate our schematizations, but also they schematize by themselves. As De Mul points out in this book, whereas in old technologies such as writing the products of thinking are outsourced to an external memory, in the case of digital technologies thinking itself is outsourced to an external device. In other words, digital technologies can be said to be, as I have already written in a previous article (Romele 2018), "imaginative machines."

However, before presenting what I mean by imaginative machines and *e*magination, it remains to be seen what I mean by imagination, tout court. The history of philosophy has been marked by an alternative between two notions of imagination, which is somehow exemplified by the tension in Kant between imagination as described in the first and in the third *Critique*, respectively. In the *Critique of Pure Reason*, productive imagination appears as the mediating link between the passivity of impressions and the activity of the judgment of concepts. It is, then, a cognitive function that simplifies and articulates sense data, giving them a unity, and hence a meaning. It is schematization. In the *Critique of Judgement*, instead, imagination does not schematize (or at least it does not schematize properly; that is, according to the concept. In fact, in the *Critique of Aesthetic Judgment*, which is the first part of the third *Critique*, Kant seems still to be aware that imagination is always an interplay between spontaneity and the capacity of giving form). There is a "free play" between imagination and understanding. And imagination definitely frees itself from the concept in the aesthetic ideas, to which "no [determinate] *concept* can be adequate, so that no language can express completely and allow us to grasp it" (Kant 1987, 182). In the second part of the third *Critique*, the *Analytic of the Sublime*, and more precisely in § 49, the German philosopher speaks of an imagination that is not productive anymore, but creative, free from all law of association and nature. It is the imagination of the artist and the genius: the painter, the sculptor, and the poet.

In this context, I cannot debate the inconsistency of the clear-cut distinction, so classic in philosophy (let us think back to Bachelard), between the kinds of imagination at work in technoscience and poetry. I am restricting myself by referring the reader back to the work of Max Black (1962) on models and metaphors, or to the "puzzling cases" of semi-arts and semi-techniques such as industrial design, architecture, and data visualization. On this basis, I propose to call the two notions (productive and creative) of imagination I have just presented through Kant, "imagination-*bricolage*" and "imagination-engineering," respectively. I am clearly referring to the distinction between *bricoleur* and engineer introduced by Claude Lévi-Strauss in *The Savage Mind* (1966).

For the French anthropologist, the starting point is the "Neolithic Paradox"; that is, the long stagnation of human scientific and technological development after the great advancements of that time: pottery, weaving, agriculture, domestication of animals, and so on. According to him (Lévi-Strauss 1966, 15), "there is only one solution to the paradox, namely, that there are two distinct modes of scientific thought ... one roughly adapted to the stage of perception and the imagination; the other at a remove from it ... one very close to, the other more remote from, sensible intuition." It is precisely in order to find an analogy today and here, in Western societies, of the former kind of scientific thought, closer to perception and imagination, not "primitive" but "prior," that he introduces the notion of *bricolage*. "The *bricoleur*," Lévi-Strauss (1966, 17) explains, "is adept at performing a large number of diverse tasks; but, unlike the engineer, he does not subordinate each of them to the availability of raw materials and tools conceived and procured for the purpose of the project." Rather, he goes on, "his universe of instruments is closed and the rules of his game are always to make do with 'whatever is at hand'"; the expression in French is "*moyens du bord*," the literal meaning of which is "the means at [disposal on] board," on a boat, for instance, when the land is still far away, and one has to make do. These resources represent, he adds in the same passage, "a set of tools and materials which is always finite and is also heterogeneous because what it contains … is the contingent result of all the occasions there have been to renew and enrich the stock." When I first read this description, I immediately thought back to my father's garage, where there is no room for a car anymore, but one can find an old motor scooter, a collection of graphic novels, tools for gardening, for home maintenance and repairing, for cutting wood and metals, old shoes, new shoes for hiking, and many other things that seem to offer a solution to every domestic problem and need. My father is a *bricoleur*; I am not. And yet, as I will argue in the following lines, we have the same spirit, and the same imagination at work. For Lévi-Strauss (1966, 19), the difference between the engineer and the *bricoleur* is clear, because while the former "is always trying to make his way out of and go beyond the constraints imposed by a particular state of civilization," the latter "by inclination or necessity always remains within them." The engineer uses the concept, while the *bricoleur* resorts to the signs, whose specificity consists in being both material and conceptual, requiring then "the interposing and incorporation of a certain amount of human culture" (Lévi-Strauss 1966, 20). He also affirms, a few sentences below, that the concepts are operators "opening up," while signs and signification have to do with the mere "reorganization."

It is precisely this distinction, the existence of an engineer capable of working with many concepts and practically without words, on the one hand, and a *bricoleur* without much inventiveness, on the other hand, that has been generally problematized and in some cases sharply criticized (Mélice 2009). For instance, Derrida (1978, 285) says that "the engineer, whom Lévi-Strauss opposes to the *bricoleur*, should be the one to construct the totality of his language, syntax, and lexicon. In this sense the engineer is a myth." And he continues by affirming that "a subject who supposedly would construct it 'out of nothing,' 'out of whole cloth,' would be the creator of the verb, the verb itself. The notion of the engineer who supposedly breaks with all forms of *bricolage* is therefore a theological idea." Which means, in other words, that we as human beings all are *bricoleurs*, and that it has never been and never will be otherwise. We have never been engineers. This same statement holds true for imagination. The imagination-engineering is a myth. Like my father, I am a *bricoleur*, as far as my knowledge, my writings, and even my most brilliant intuitions are concerned. They all depend on my competences in history of philosophy and its theory. I just *bricole* with words and thoughts, rather than with things and tools. Novelty and creation are no more than a shibboleth, a minimal deviation from one recombination to another. On this point, Maurizio Ferraris (2012, 316) points out: "Now, it is this very error, this shibboleth, that picks out the individuality of individuals and characterizes their uniqueness." For him, there are two ways of explaining the uniqueness of persons. The first one is "hifalutin and solemn, and it insists on our positive exceptionalness because we are infinite and ineffable," as when one talks of geniality and creativity. The second one, and this is the way I see imagination, defines uniqueness as "a negative exceptionality, a production error so to say."

It is, I believe, the passage from exteriority, the passage we have seen in Simondon, which has been theorized by other philosophers but neglected by a long tradition, that brings us to temper our expectations towards human imagination and schematism. Such a passage, through language, culture, society, technology, body, and nature grounds imagination in the world, namely in the limits and possibilities offered to us by the reality itself.

3   Automatic Aesthetics

And yet, we *see* novelty, we *see* creativity, and we *see* originality, especially in fields such as the art world. Interestingly enough, Lévi-Strauss collocates the artist halfway between the engineer and the *bricoleur*, because "by his craftsmanship he constructs a material object which is also an object of knowledge" (Lévi-Strauss 1966, 22). Which means that the artist is the one who does or provokes thoughts via things: colors, brushes, canvas, and so on. However, this is still a simplistic way to present the *bricolage* of the artist. There is, according to me, a deeper way that has to do with the very essence of the art world. Let us think of an artwork, a sculpture representing the US President Donald Trump struck down by a meteorite on a red carpet, and shards of glass next to his body. And suppose two visitors at the gallery where this piece is exhibited, Alice and Arthur. Alice does not know much about art, but she is very interested and engaged in politics. She is particularly horrified about Trump's idea of building a wall along the USA-Mexico border, one of the famous central arguments of his presidential campaign. She also recently read in *Vanity Fair* that President Trump would like this wall to be "transparent," because drug dealers may otherwise throw large bags of drugs across to the other side, and hit innocent passers-by.[1] When she sees the wax sculpture in the gallery, the meteorite, and the pieces of glass, she is immediately reminded of the "transparent" wall, and of the 60lb bags thrown by drug dealers. She smiles, she thinks this artwork is provocative, she sees political criticism and satire in it. She identifies with the intentions of the author. In sum, she loves it.

Arthur is not much interested in politics. Of course, he knows about the wall, and he has heard from his friend Alice about the idea of a "transparent" wall, but he is rather resigned to the stupidity of human beings as political animals. For this reason, for some years now, he has sought refuge in the study of art history and theory. He also loves to frequent art galleries and exhibitions. Last year, he went to Paris, and he visited the Italian artist Maurizio Cattelan's exhibition, *Not Afraid of Love*, at the Monnaie de Paris. Arthur had not been enthusiastic about art in 2011, when Cattelan had his retrospective at the Guggenheim in New York. In Paris, he finally saw the much discussed installation *La nona ora* (*The Ninth Hour*), a life-sized effigy of Pope John Paul II hit by a meteor, on a red carpet, with pieces of glass next to him, that were first exhibited in 1999 at the Kunsthalle Basel. He loved that work, its title and the multiplicity of its possible interpretations. And of course, he did not appreciate the Trump statue, which is for him nothing but a bad imitation of Cattelan's installation. At the very least, the artist could have replaced the meteorite with a large bag, and the red carpet with the dry field that characterizes many parts of the USA-Mexico border. Contemporary art and culture is essentially "quotationist," but one has to be careful about the fine line between quotation and plagiarism.

With this example I want to show how the appreciation of an artwork depends on the competence we have concerning the art world, which is essentially made of art history and theory, and to a certain degree of society: institutions, critics, galleries, and so on. At least, this is Arthur Danto's thesis in his major work, *The Transfiguration of the Commonplace* (1981). My version of this thesis is that the more expertise we gain when it comes to the art world, the more we see the recombinatory nature of a piece of art. This also means that the more we know the art world, the less we have the illusion of pure novelty and creation. Finally, this also implies that the more confidence we have regarding the art world, the more we are able to discriminate between the repetitive and innovative recombinations. Such thesis can be applied to both the aesthetic judgment and the aesthetic production, since art, as Danto points out, is henceforth (or, it has always somehow been) intellectualized, which means that an artist such as Cattelan does not work in isolation, but always in dialogue with the past and the present of the art world.

You probably have already guessed my strategy. Until now, I have just spent a few words on digital technologies and media, and I have focused my attention on human imagination. Instead of exalting digital imagination, I have slightly criticized the human imagination. My inspiring figure has been Bourdieu, and the illuminating pages he wrote against Sartre, describing the *habitus* in *Outline of a Theory of Practice* (1977). I am perfectly conscious of the risk of appearing a determinist, precisely as Bourdieu has been accused, of being partly with good reason. I will discuss this point in the conclusion.

It is time now to spend some words on digital technologies. In a previous article of mine that I have already quoted (Romele 2018), I resorted to Lev Manovich's notion of software to show how digital technologies can be understood as imaginative machines, in the sense of productive imagination. According to him, the software is the basis of all digital expressions, and "if we don't address software itself, we are in danger of always dealing with its effects rather than the causes" (Manovich 2013, 9). The logic of software relies for Manovich (2013, 207) on the articulation of two elements, namely databases and algorithms: "a medium as simulated in software is the combination of a data structure and set of algorithms [. . .]. We have arrived at a definition of a software "medium," which can be written in this way: Medium = algorithms + a data structure."[2] And a few pages below, he proposes this interesting twofold analogy: "To make an analogy with language, we can compare data structures to nouns and algorithms to verbs. To make an analogy with logic, we can compare them to subjects and predicates" (Manovich 2013, 211). In that article, I said that to make an analogy with the productive imagination, one could compare databases to sensibility and algorithms to the forms of understanding. The function of an algorithm consists of reorganizing data according to a certain coherence. Examples are an image on Instagram as a collection of pixels on which different filters are applied; an Excel document on which we can apply different algorithms for data treatment and visualization; from a user perspective, the entire web is a database on which algorithms like Google PageRank operate.

It is precisely within this material duality that, I believe, one has to understand digital recording and registration. And this is also the reason why, in several chapters of this book, authors have insisted on the fact that digital registration and recording are not passive, but rather imply activity and agency. There is, however, an important difference between *manually* applying a filter on an image on Instagram, or *manually* entering a search on the Google search engine for some news, and leaving the machine to *automatically* look up the information for us. Galit Wellner has showed it in her chapter, by discussing "algorithmic writing." An article recently published on *Wired* speaks about Google's new algorithm that "perfects photos before you even take them."[3] Even without going that far, one can consider Google Photos' assistant, which autonomously proposes stylized photos, animation from videos, and so on, or Google News that suggests news that is supposed to be relevant to you. And one might go on about hundreds of services of this kind offered by Google and other enterprises.

The question arises as to what extent can we say that we are before a creative imagination? On the basis of what I have said before, novelty and creation are mostly observer-dependent. This holds true for both human beings and machines. For example, the first time one sees Facebook's Year in Review video, one will probably have the impression of ad hoc video editing. The reiterated observation and the confrontation with other users will reveal, however, its determinism.

Certainly, the complexity of the object observed plays an important role. Before it became unplayable, I amused myself with the video game designed by Stefano Gualeni *Gua-Le-Ni; or, The Horrendous Parade*, based on Hume's notion of complex ideas. *Gua-Le-Ni* took place on the wooden desk of an old British taxonomist. On his desk lay a fantastic book: a bestiary populated by finely drawn creatures. As for the monsters of myths and folklores in general, the impossible creatures in *Gua-Le-Ni* were combinations of parts of real animals. The goal of *Gua-Le-Ni* was that of recognizing the modular components of the fantastic creatures and their relative order before one of them managed to flee from the page.[4] Now, let us imagine that the possible combinations of *Gua-Le-Ni* were not 30, but thousands or even millions, as numerous and small as the pixels of a smartphone's good camera are. We would not be able to play the game, because we would be unable to recognize the modular components of the fantastic creatures. And we would not recognize that the impossible creatures we see are the result of recombinations of parts of other creatures. They would appear as new creatures. One could certainly say that such recombinations would result in mere chaos. But let us suppose that at the core of *Gua-Le-Ni* is a machine learning algorithm that has been trained with images of what we commonly consider as creatures. The algorithm would then create images that are similar, but not equal, to other creatures we have seen before. We would have brand-new images of fantastic creatures, without any perception of the determinism that lies behind them. Interestingly enough, this example is complex and maybe inappropriate not because of the limits of the machines, but because it questions the limits of our imagination. Can we as human beings really imagine brand-new creatures? Rather, is not our imagination doomed to remain within the limits of anthropomorphism or at least of what has been already seen on Earth? The machine must be fed with our standards, but this not for the machine's own sake, but rather for our needs and fears. The recent case of two of Facebook's AI programs is very well known. The programs were shut down because they appeared to be chatting to each other in a language only they understood.[5]

In a recent article from which I have borrowed the title of this section, Manovich (2017) argues that artificial intelligence is playing an important role in our cultural lives, increasingly automating the realm of aesthetics. In particular, he insists on the fact that today AI is not just automating our aesthetic *judgments*, recommending what we should watch, listen to, or wear. It also plays an important role in some areas of aesthetic *production*. AI is used to design fashion, logos, music, TV commercials and other products of the culture industry (Manovich 2017, 3). For sure, in most cases, it is still fair to speak of a data-*driven* culture, in the sense that human beings still make the final decision. For example, this is the case with *Game of Thrones*, in which the computer suggests plot ideas but the actual writing is done by humans. This is also the case of what has been called the first "AI-made movie trailer," that of the sci-fi thriller *Morgan*, made with the help of IBM Watson.[6] The computer selected many shots suitable to be included in the trailer, and then a human editor made the work. But it is only a matter of time before entirely AI-designed cultural products will be available on the market of the culture industry; and actually, they are already widely available in the art world.[7] Manovich (2017, 6) also stresses how the use of AI and other computer methods is helping us find regular patterns in culture, such as influence between artists, changes in popular music, and gradual changes in average shot duration across thousands of movies. This means that digital machines are teaching us to be modest when it comes to us pretending to be engineers.

4   Conclusion

The idea which forms the basis of this chapter can be formulated as follows: one must certainly recognize that some digital machines, such as the algorithms of machine learning, especially when applied to specific fields like culture, are manifesting emerging properties that were considered human peculiarities until some years ago. Yet, one should also consider the possibility that this is happening not only because some of these machines are extraordinary, but also because human beings are revealing themselves to be more ordinary than what we have been usually disposed to believe. Even when it is a matter of invention, creativity, and imagination. In other words, the aim of this chapter has not been so much to glorify digital machines (although I believe there are several good reasons to be stupefied by some of their most recent performances), but to reduce our self-esteem a little.

The main advantage of this approach is that I do not have to position myself among the techno-enthusiasts or the transhumanists in order to defend my principle of symmetry. The biggest disadvantage, however, is to appear, as I have already said, a determinist. I have no difficulty in recognizing myself as a defender of social determinism. In this case, I would even define myself as a staunch defender of determinism, as staunch as the Foucault of the first volume of *The History of Sexuality*, in which he describes the process of interiorization of the obligation to confess (*avouer*): "The obligation to confess is now … so deeply ingrained (*incorporée*) in us, that we no longer perceive it as the effect of a power that constrains us; on the contrary, it seems to us that truth, lodged in our most secret nature, 'demands' only to surface"; we believe that "truth does not belong to the order of power, but shares an original affinity with freedom" (Foucault 1978, 60). In social life, freedom is no more than a vanishing point. The same holds true, for instance, for the creation of an artwork if seen from a specific art world perspective *à la* Danto. Neither do I have difficulty in recognizing myself as a determinist whenever it is a matter of mediation and exteriorization, such as in technology, language, and embodiment.

And yet, I have some difficulties in affirming that determinism is the ultimate human condition. Maybe this is just a case of residual human conceit. And maybe from a higher perspective, that of a god or a superior machine, we would appear as pure matter, with all our behaviors already inscribed in our (genetic?) code. All I know is that as our knowledge stands at present, we can still *believe* in a tension, although minimal, between auto- and hetero-determination. This is not, however, the case with digital machines, even the most incredible ones, because however complex and black-boxed, we *know* they are no more than algorithms and databases. It is precisely for this reason that we are henceforth disposed to attribute a certain morality to technologies, but it would be rather odd to discuss their responsibilities (Floridi and Sanders 2004).

Notes

1. Maya Kosoff, "Trump Ants a "Transparent" Border Wall to Prevent Injuries from Falling "Sacks of Drug"." *Vanity Fair*, July 13, 2017, https://www.vanityfair.com/news/2017/07/trump-transparent-border-wall-falling-drugs-mexico
2. There is actually a difference between database and data structure, and this is the reason why Manovich remains, in his definition, a sort of "material idealist." I thank Galit Wellner for this precious remark.
3. Elisabeth Stinson, "Google's New Algorithm Perfects Photos Before You Even Take Them." *Wired*, July 8, 2017, https://www.wired.com/story/googles-new-algorithm-perfects-photos-before-you-even-take-them/
4. "Gua-Le-Ni; or, The Horrendous Parade." *Wikipedia*, https://en.wikipedia.org/wiki/Gua-Le-Ni;_or,_The_Horrendous_Parade. Accessed October 2, 2017.
5. Andrew Griffin, "Facebook's Artificial Intelligence Robots Shut Down Because They Start Talking to Each Other in Their Own Language." *Independent*, July 31, 2017, http://www.independent.co.uk/life-style/gadgets-and-tech/news/facebook-artificial-intelligence-ai-chatbot-new-language-research-openai-google-a7869706.html
6. "Morgan. IBM Creates First Movie Trailer by AI." *YouTube*, https://www.youtube.com/watch?v=gJEzuYynaiw. Accessed October 2, 2017.
7. See, for instance, Issue 39 (2015/2016) of the French review *artpress2*, devoted to the digital arts.

References

Bachimont, Bruno. 2010. *Le sens de la technique. Le numérique et le calcul*. Paris: Les belles lettres.

Beaubois, Vincent. 2015. Un schématisme pratique de la raison. *Appareil* 18. https://appareil.revues.org/2247?lang=en

Black, Max. 1962. *Models and Metaphors*. Ithaca, NY: Cornell University Press.

Bourdieu, Pierre. 1977. *Outline of a Theory of Practice*. Cambridge: Cambridge University Press.

Danto, Arthur. 1981. *The Transfiguration of the Commonplace. A Philosophy of Art*. Cambridge, MA: Harvard University Press.

Derrida, Jacques. 1978. *Writing and Difference*. Chicago: The University of Chicago Press.

Desrosières, Alain. 2002. *The Politics of Large Numbers. A History of Statistical Reasoning*. Cambridge, MA: Harvard University Press.

Domingos, Pedro. 2015. *The Master Algorithm*. New York: Basic Books.

Estivals, Robert. 2003. *Théorie générale de la schématisation 3. Théorie de la communication*. Paris: L'Harmattan.

Ferraris, Maurizio. 2012. *Documentality. Why It is Necessary to Leave Traces*. New York: Fordham University Press.

Floridi, Luciano, and James Sanders. 2004. On the Morality of Artificial Agents. *Minds and Machines* 14 (3): 349–379.

Foucault, Michel. 1978. *The History of Sexuality. Volume 1: An Introduction*. New York: Pantheon Books.

Ginzburg, Carlo. 1989. Clues. Roots of an Evidential Paradigm. In *Clues, Myths, and the Historical Method*, 96–125. Baltimore: John Hopkins University Press.

Kant, Immanuel. 1987. *Critique of Judgment*. Indianapolis and Cambridge: Hackett Publishing Company.

Latour, Bruno. 2007. Beware, Your Imagination Leaves Digital Traces. *Times Higher Literary Supplement*, April 6, 2007. http://www.bruno-latour.fr/sites/default/files/P-129-THES-GB.pdf

Lenk, Hans. 1995. *Schemaspiel: über Schemainterpretation und Interpretationskonstrukte*. Frankfurt am Main: Suhrkamp.

Lévi-Strauss, Claude. 1966. *The Savage Mind*. London: Wiedenfeld and Nicolson.

Maddalena, Giovanni. 2015. *Philosophy of Gesture. Completing Pragmatists' Incomplete Revolution*. Montreal and Kingston: McGill-Queen's University Press.

Makkreel, Rudolf. 1990. *Imagination and Interpretation in Kant. The Hermeneutical Import of the Critique of Judgment*. Chicago: The University of Chicago Press.

Manovich, Lev. 2013. *Software Takes Command*. London: Bloomsbury.

———. 2017. Automatic Aesthetics. Artificial Intelligence and Image Culture, 2017. Accessed September 12, 2017. http://manovich.net/content/04-projects/101-automating-aesthetics-artificial-intelligence-and-image-culture/automating_aesthetics.pdf

Mélice, Anne. 2009. Un concept lévi-straussien déconstruit: le "bricolage". *Les temps modernes* 5 (265): 83–98.

Romele, Alberto. 2018. Imaginative Machines. *Techné: Research in Philosophy and Technology* 22 (1): 98–125.

Simondon, Gilbert. 1989. *Du Mode d'existence des objets techniques*. Paris: Aubier.

———. 2014. *Imagination et invention (1965–1966)*. Paris: Presses Universitaires de France.

Index

© The Author(s) 2018
A. Romele, E. Terrone (eds.), *Towards a Philosophy of Digital Media*,
https://doi.org/10.1007/978-3-319-75759-9

CPSIA information can be obtained
at www.ICGtesting.com
Printed in the USA
LVOW13*2359050818
586026LV00010B/529/P